DAVID MAISEL

HISTORY'S SHADOW

HISTORY'S SHADOW

JONATHAN LETHEM

X, CURATOR

DAVID MAISEL

TRACE ELEMENTS AND CORE SAMPLES

NAZRAELI PRESS

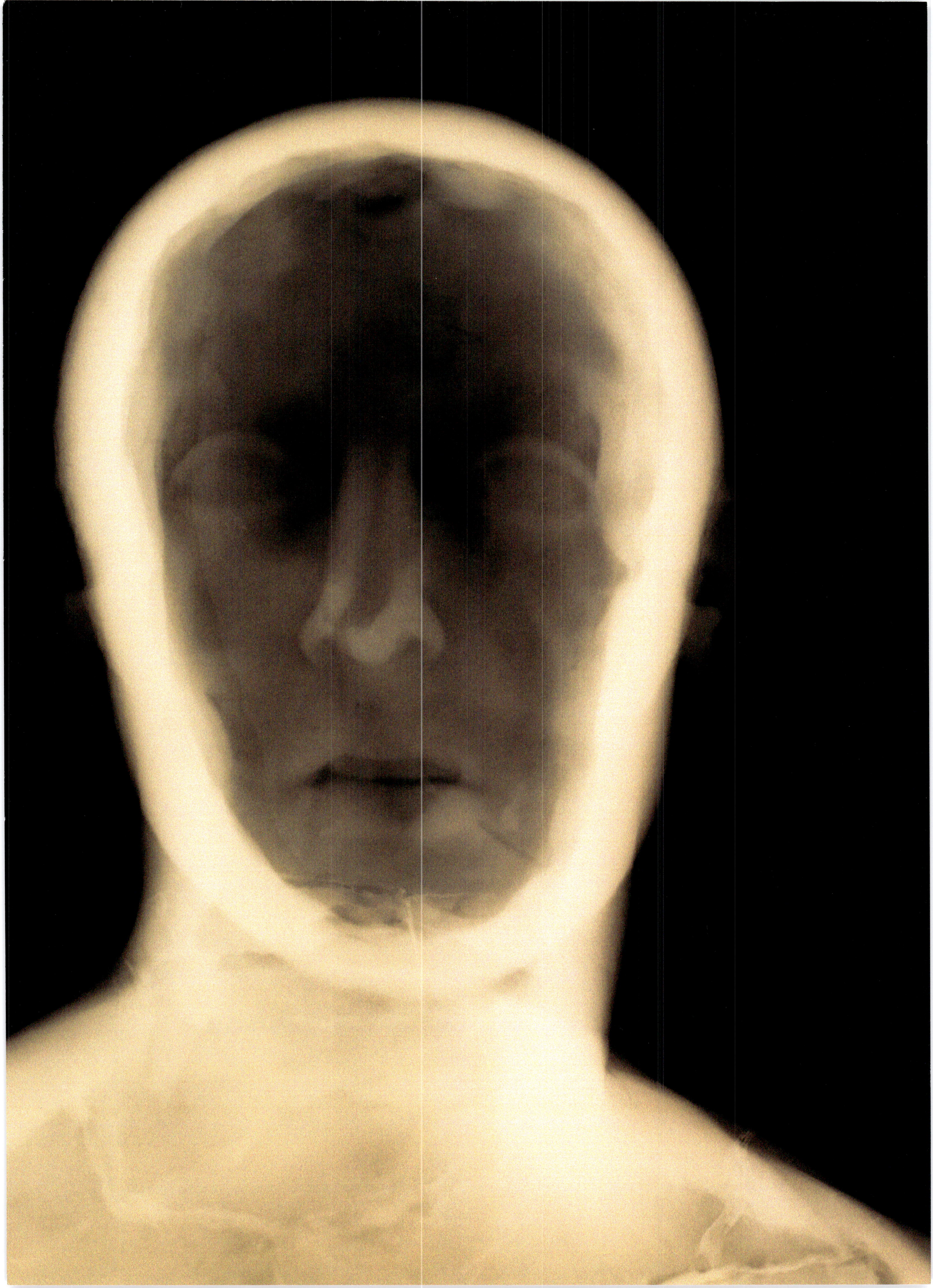

GM16 / AB15

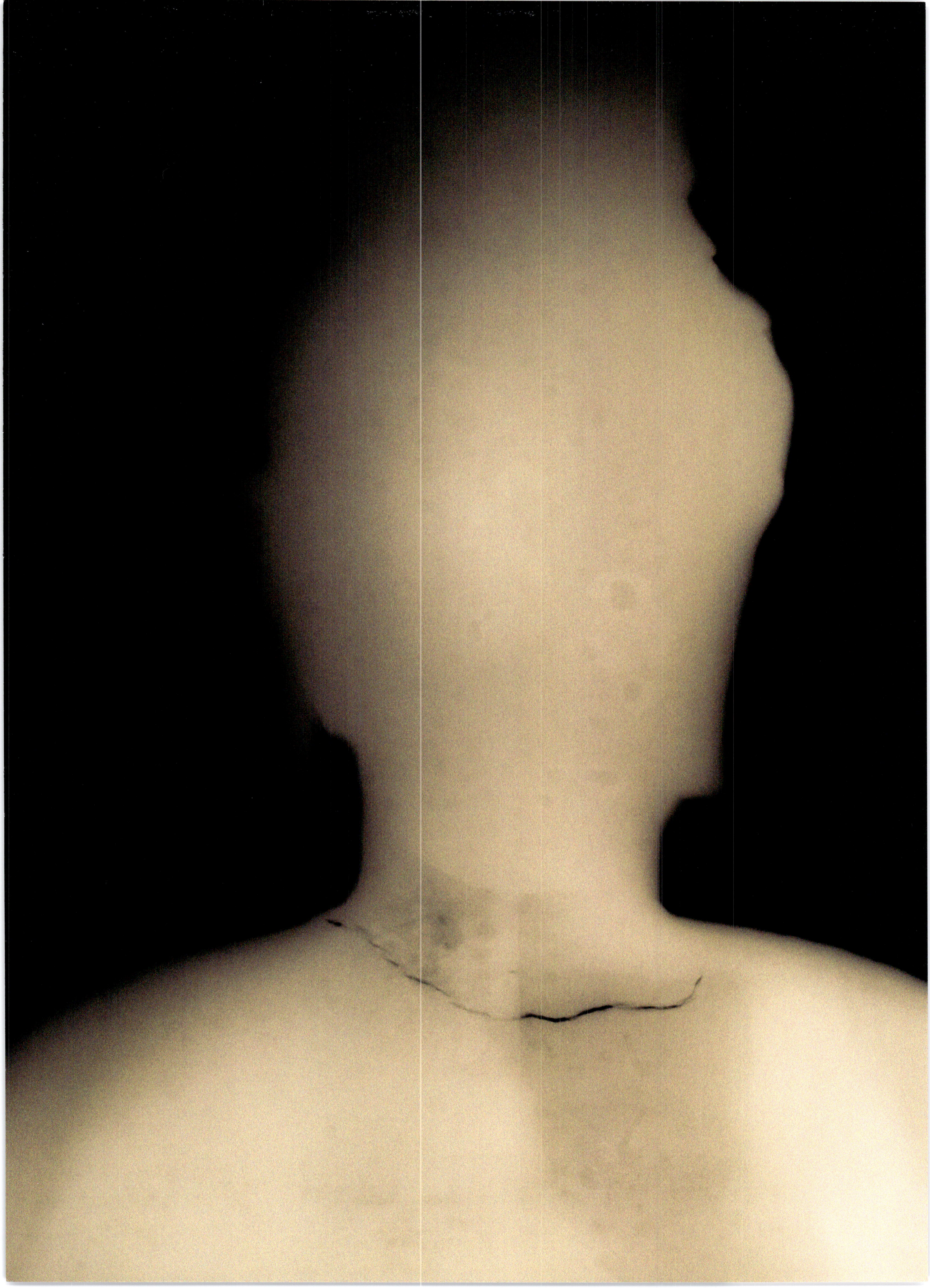

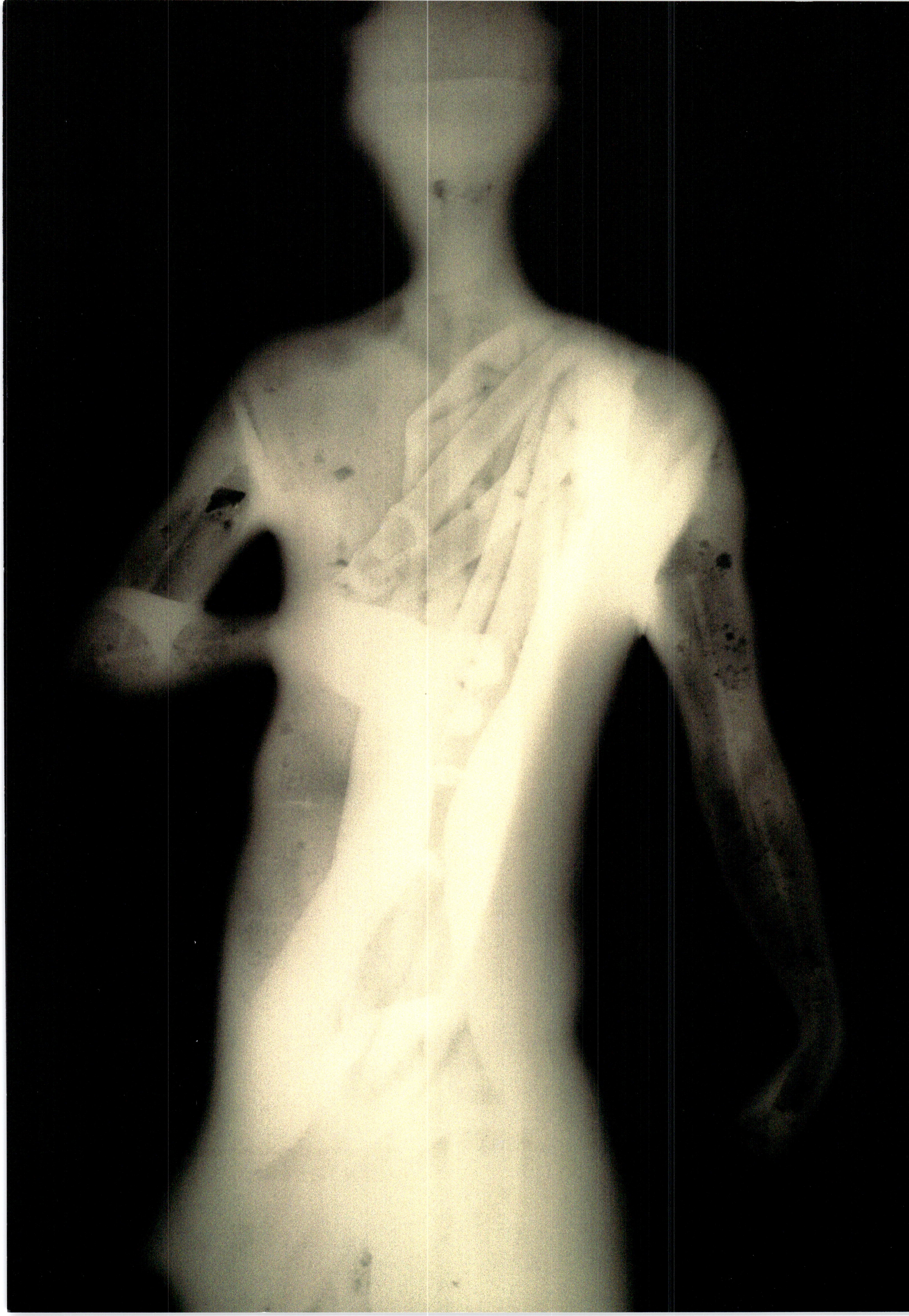

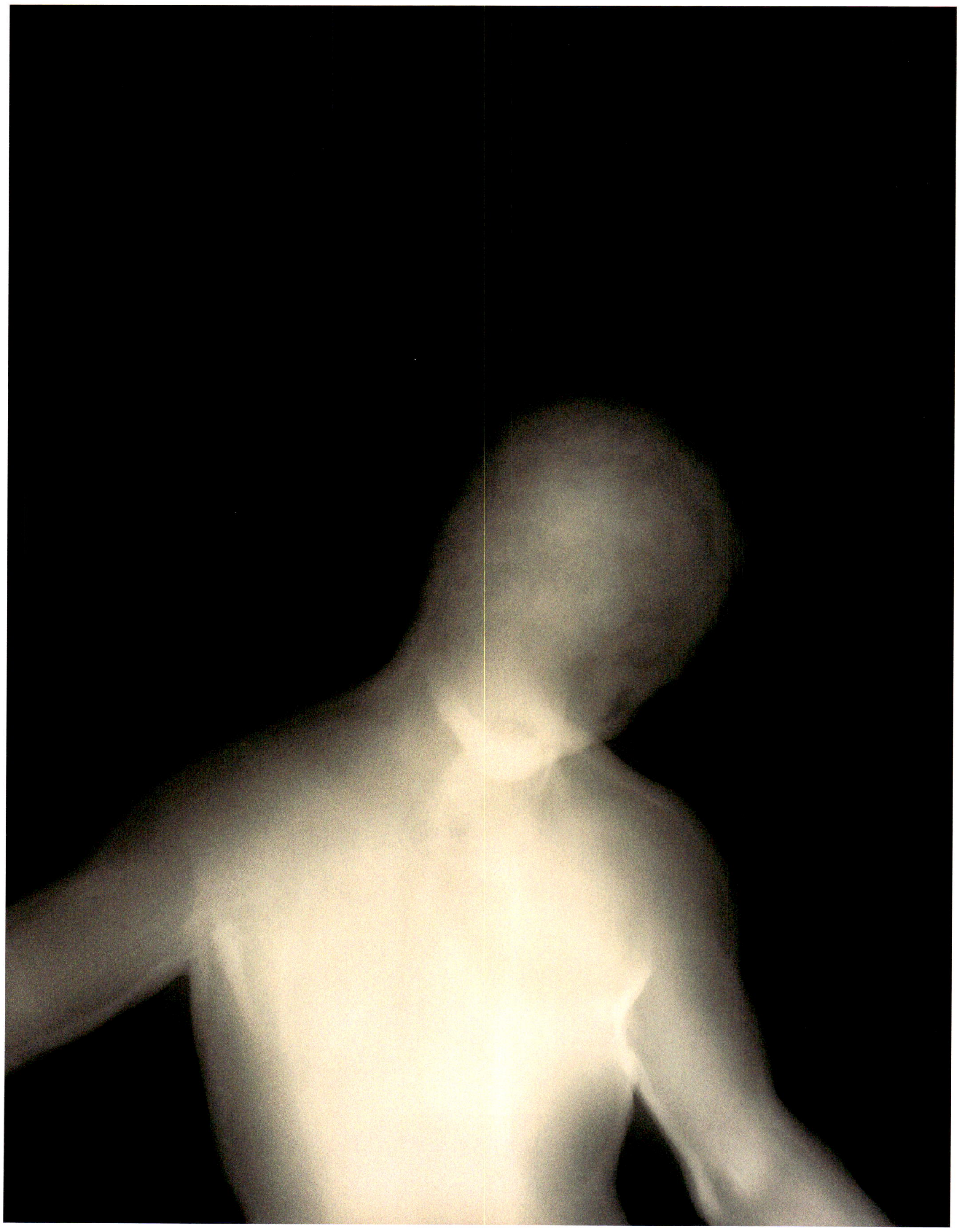

GM4 / GM19 / AB9

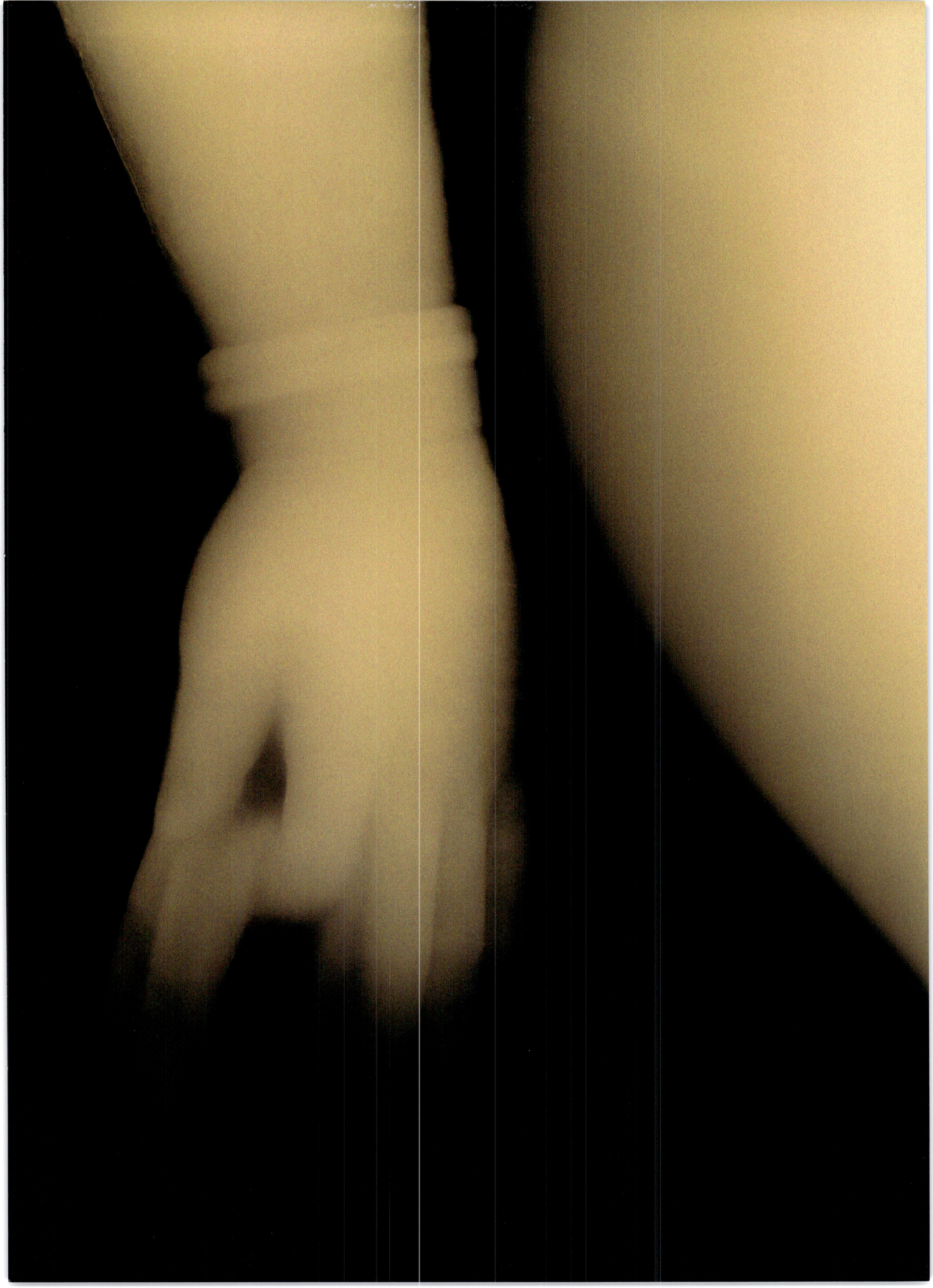

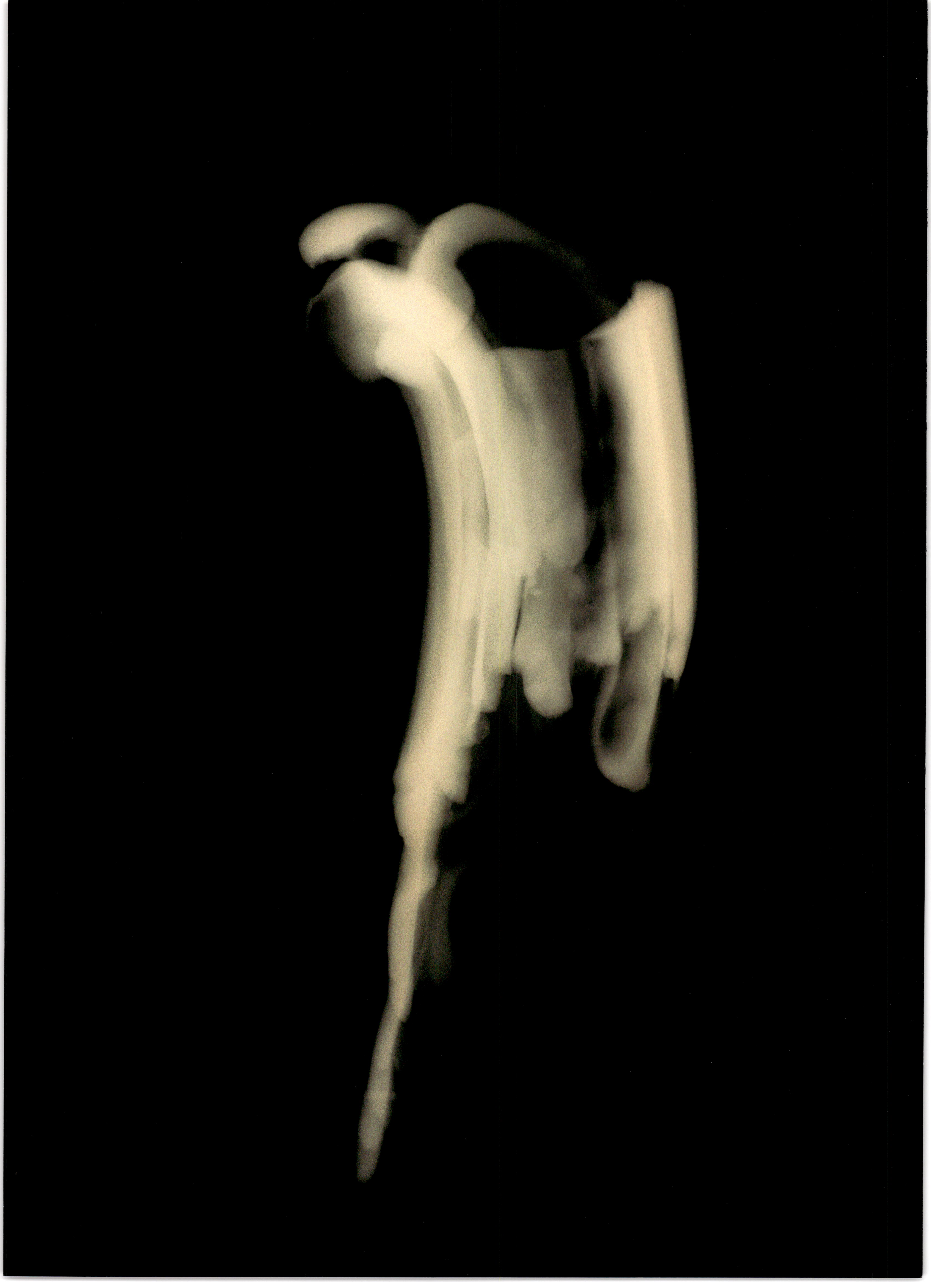

GM11

X, CURATOR

X had been traveling for what seemed centuries. He might be decades out of touch with the home office, not that he could ever have lost grasp of his purposes. The longer he ran out his assignment, the more purified the relationship between him and the objects of his pursuit. The world more and more came down just to him and those enigmatic relics he sought, the isolate treasures that delivered their essences only to his special gaze, those on which he made report and with which he sometimes abruptly absconded—assessor turned master thief at the drop of a hat—and which were otherwise arrayed as if part of the ordinary universe of artifacts. X barely noticed anything but the relics anymore. The rest was darkness, the inky milk of time. The human world legible only by implication.

X's work, the imperative of his visions, was enough to sustain him. Rarely even did a human voice pierce the caul of his obsession with the objects; a flight attendant asking if he'd change his seat so that members of a family could sit together, a desk clerk explaining that the strip on his credit card had become demagnetized: did he have another? X looked clean through these phantasms, no longer able to shorten his gaze to halt at the surfaces of bodies, unable to quit penetrating to the drear miniscule resonances already accumulating in a fountain pen or a keychain, or in a photocopy machine on a shelf behind the person speaking. Objects spoke more than persons now. In a thousand years, should these stray artifacts endure, they might have accumulated enough cosmic weight to be worthy of X's glance, even for being gathered up in the shadowy archives of his unseen handlers. For now, these fledgling objects were distractions, static on his screen. Somewhere, in this city where X had arrived, some singular statue or urn beckoned to him, radioactive with presence, metastatic with abandoned meaning. Not only did X possess a genuine calling, but if he was silent enough he could actually hear it, an involuntary whisper of inferences issued several centuries before. For X, the seer as bureaucrat, these echoes were his bona fides.

. . . .

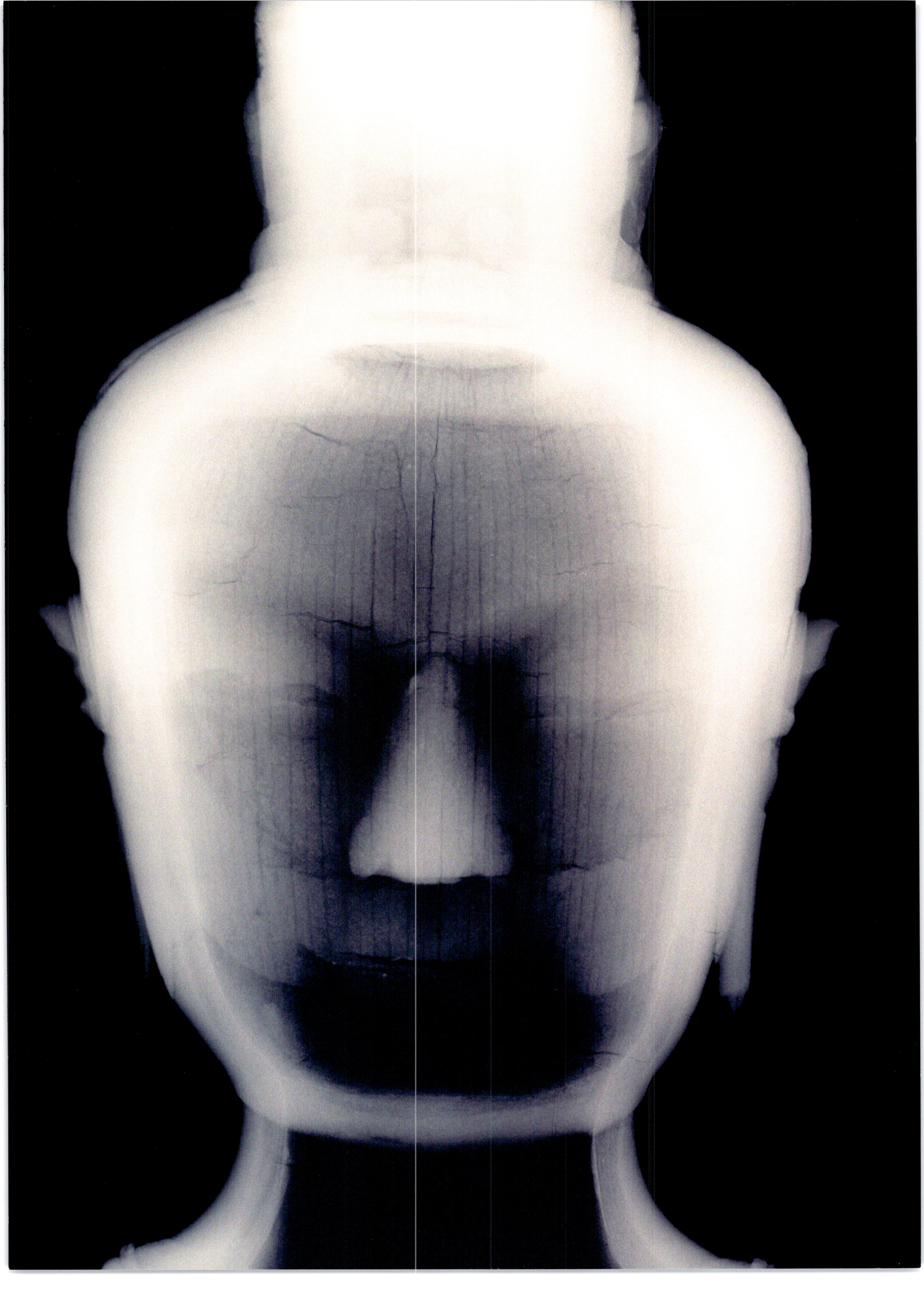

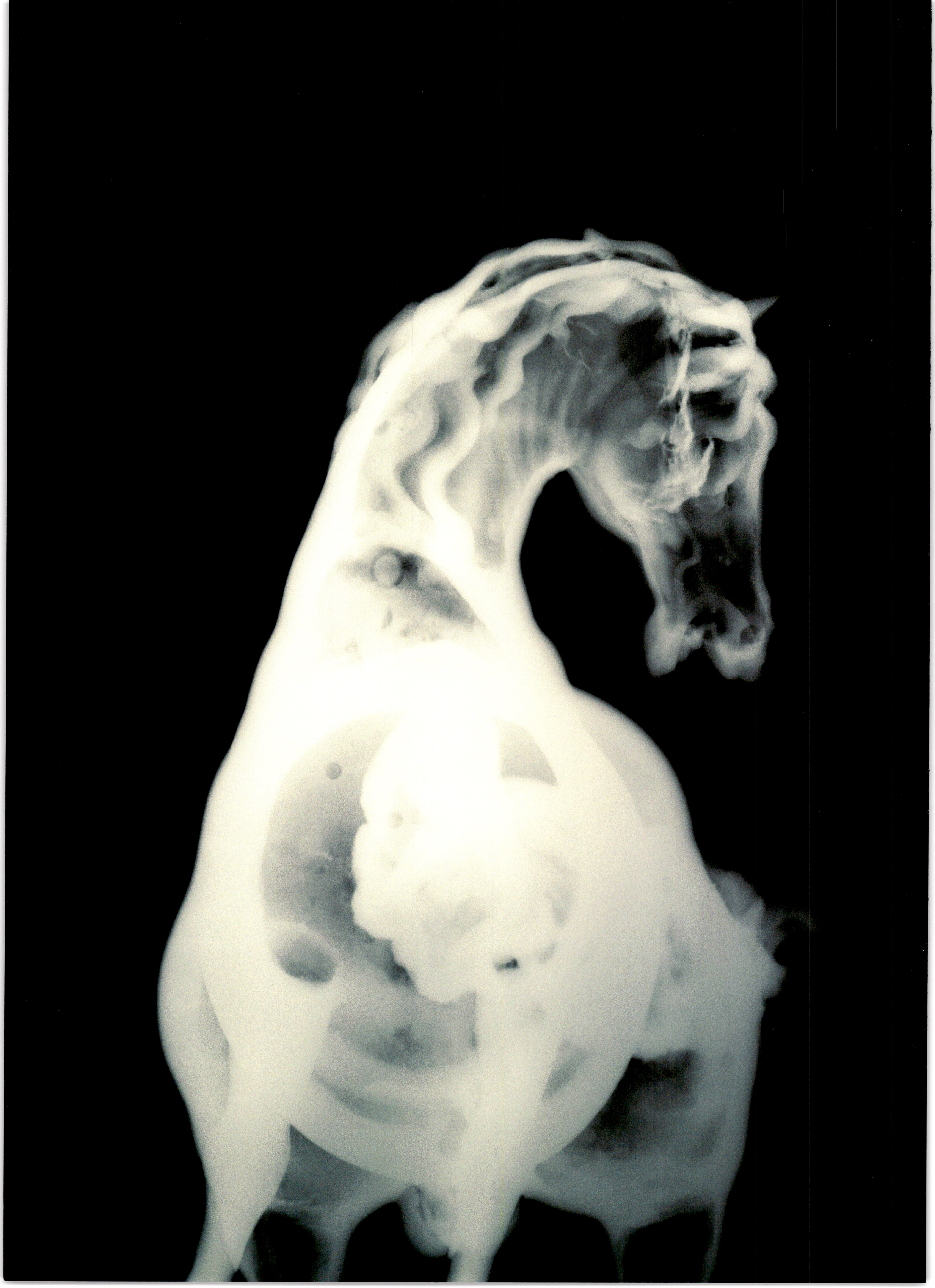

GM3 / AB6

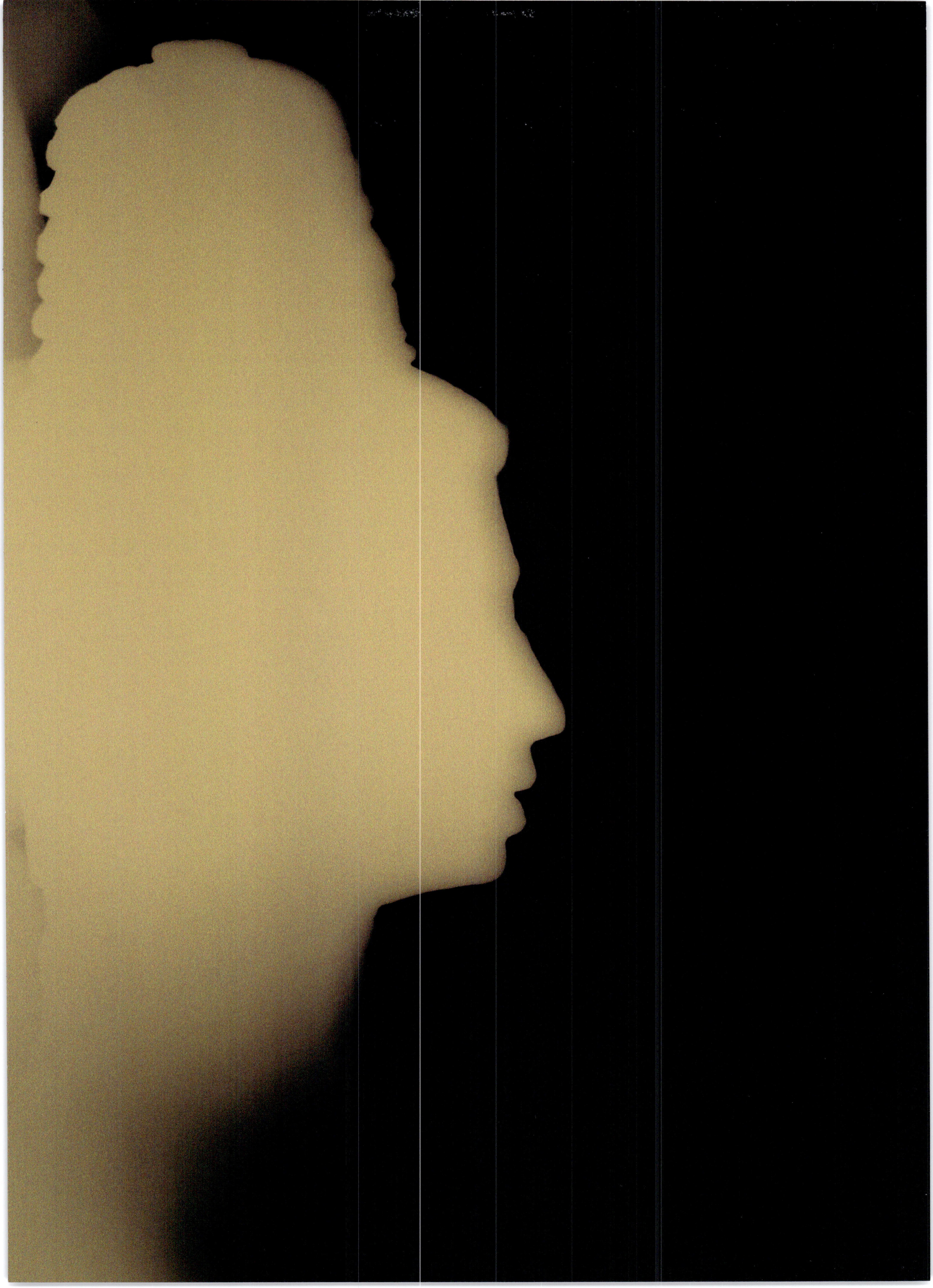

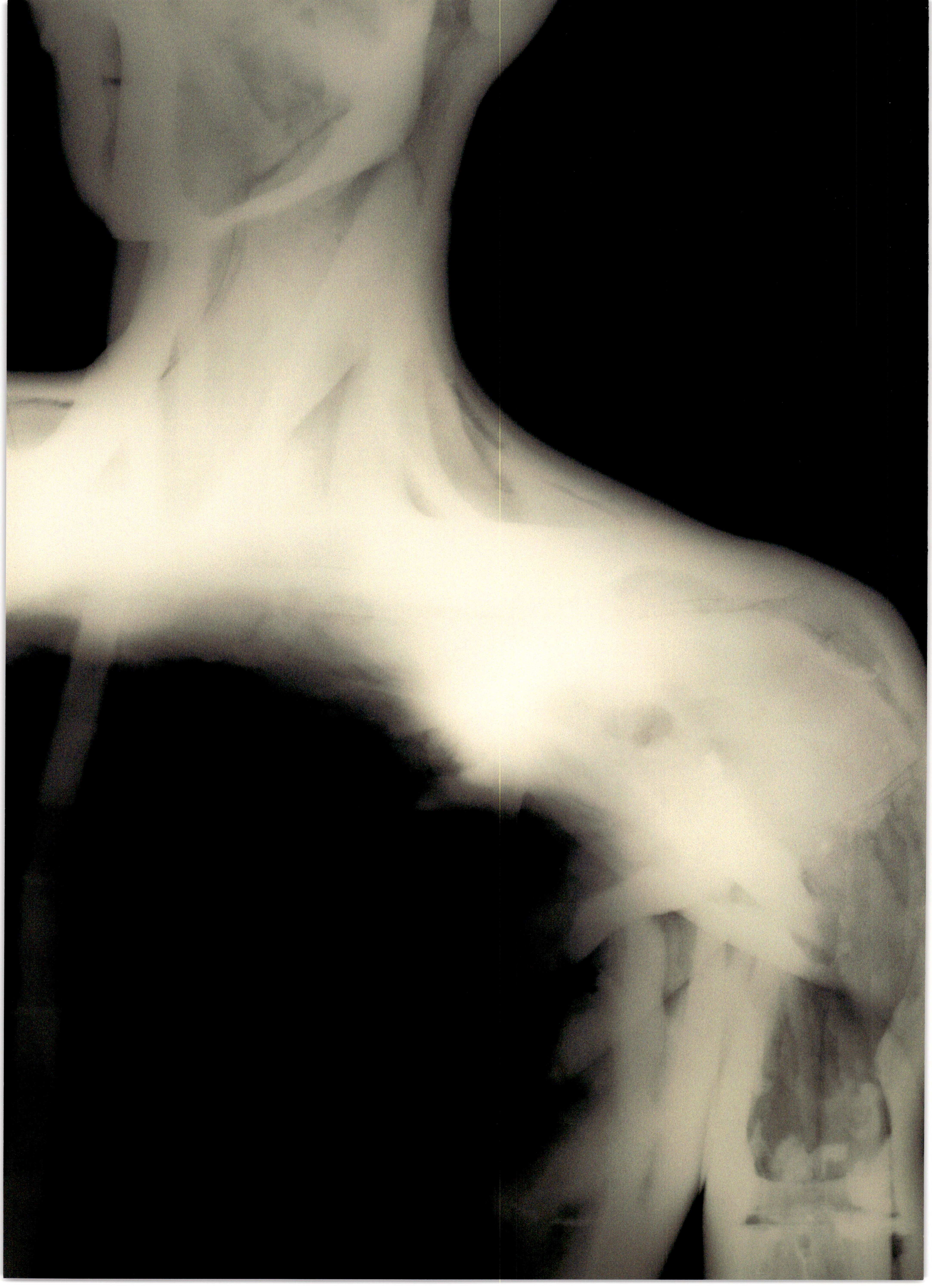

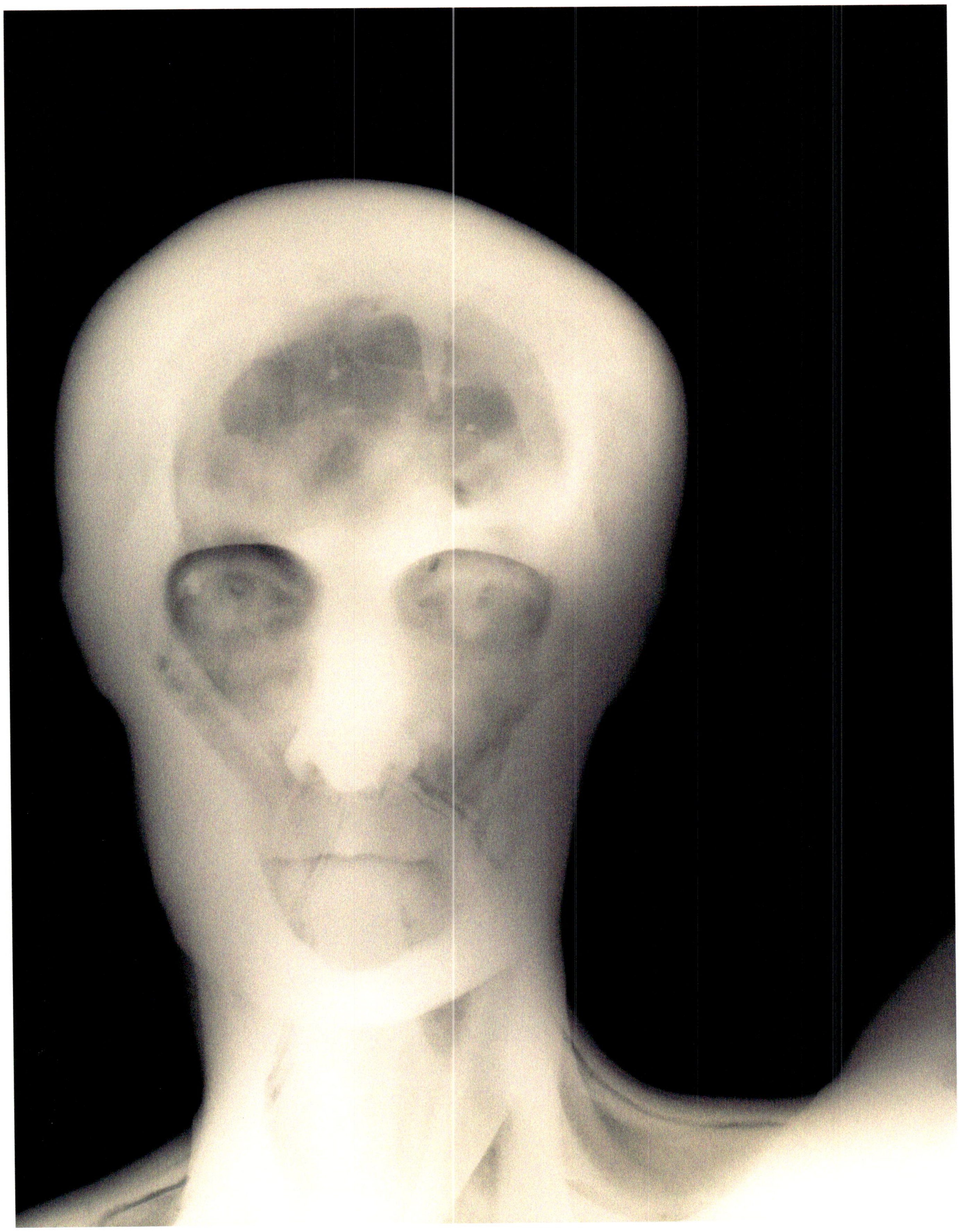

GM9 / GM1

The true conundrum, if X allowed himself to face it, was the point of origin of his assignment, and the nature of his employers. X had several theories he auditioned, during those moments of weakness when he was unable to keep speculation at bay, frequently on airplanes, with his eyes closed.

Possibly X served at the pleasure of a vast and dispassionate organization, a bureaucracy of curators at desks who filtered reports from curators in the field, and for many or even all of whom X's special advantages might be a rumor or forgotten, his talents never particularly cherished by the higher-ups in the organization, his conscription to their cause only grudging at best, his dossier gathering dust in Human Resources. The image that went with this possibility was of a pentagonal or hexagonal building, adrift in green fields, and with as great a portion of its mass underground as above, if not in fact a much greater portion, the building itself a metonym for the hidden depths in the objects the organization studied and hoarded. X a sporadic visitor to this Kafkaesque castle of indifference, until he'd come to understand that his absences went unremarked and unrecorded. Thenceforth he'd become self-authorizing, a free-ranger, letting his (ignored) dispatches stand in for his being.

The appeal and terror of this prospect was simple. X could slip through the world a solo operator, rendered, by the bigness and stupidity of a governmental-style task force, utterly alone. Under contract in some technical sense, yet answerable to no one in the end but himself. Unseen in his seeing. As isolated in the white noise of commercial travelers as his treasures were alone in their crowd of unvibrant object-companions.

Another possibility: X worked for an alien, god, or monster. Was, like the Silver Surfer employed by Galactus, a local emissary or appointed herald for some gigantic presence or force arrived on Earth from elsewhere, and whose indifference to anything but the treasures at hand—that is, of course, to the transparent persons flitting around the edges of such treasure—had invested itself over time into X's own attitudes toward the human species he still, rightly or wrongly, counted himself among.

X might merely be an extraterrestrial collector's eyes. The remote-sensing device of a treasure-eater from Mars or some further place. Perhaps ludicrous, this theory had the advantage of explaining the fact of his powers, his superseeing capacity for peering inside the skin of the world. As well as the fundamental alienation bestowed with such responsibilities.

. . . .

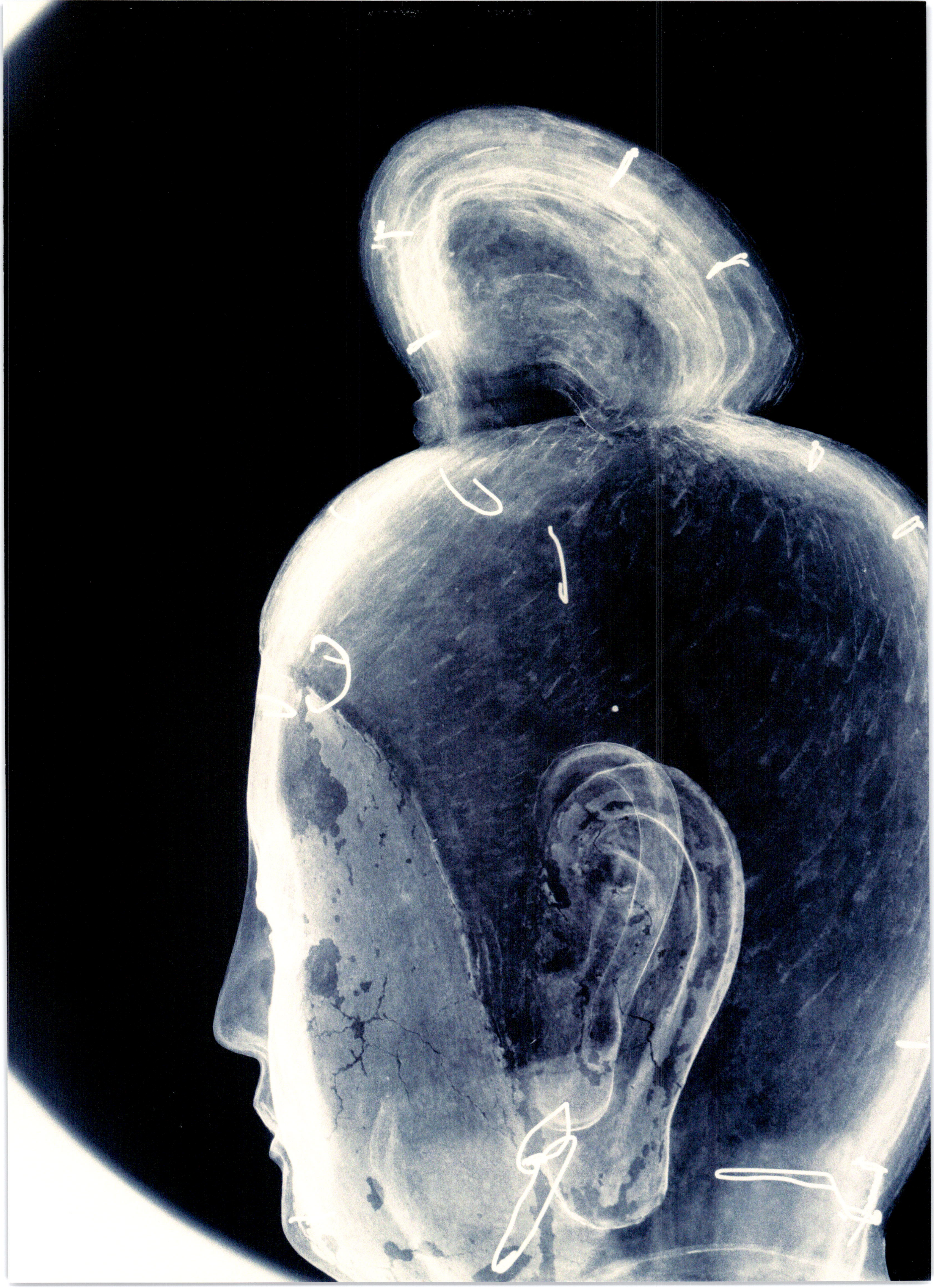

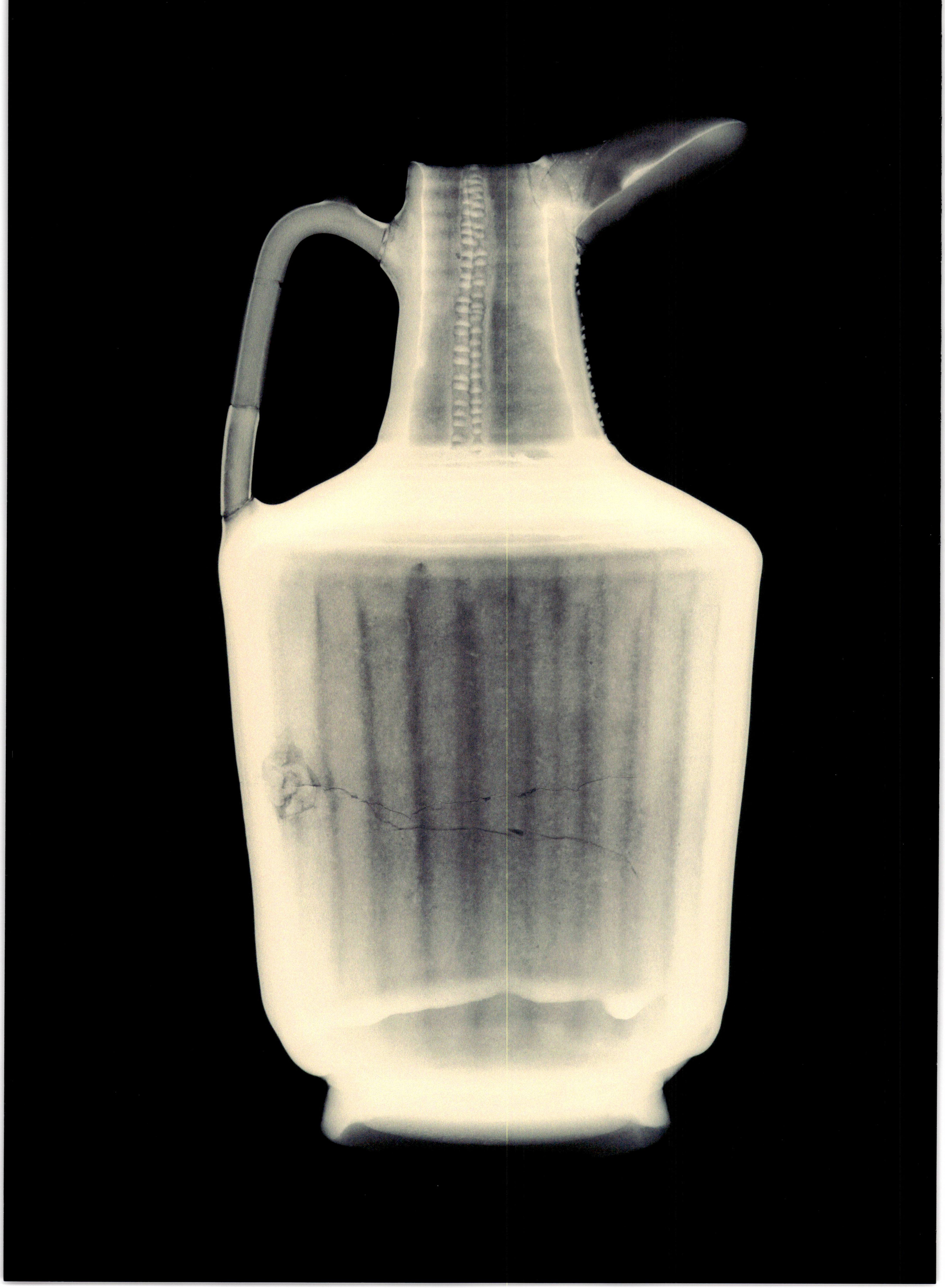

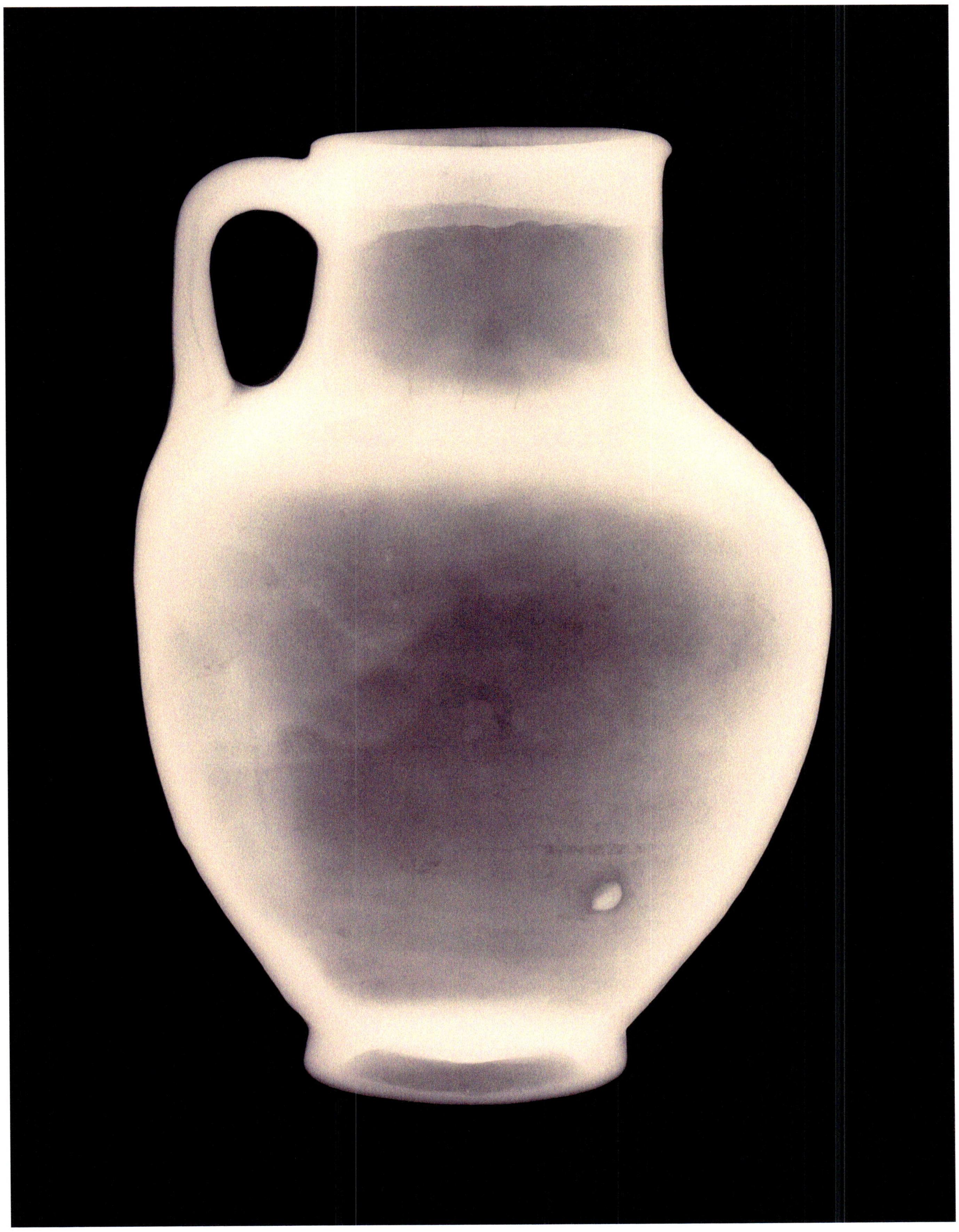

AV4 / AV1

Another, related conjecture was that X was employed by the objects themselves.

In this scenario, the treasure was a kind of giant collective mind, disassembled and decommissioned over the human centuries, but now longing for communion and reconnection. The unvocal yearning chorus, trapped in unambulatory bodies, jailed in mansions or museums, repositories of helpless cultural plunder or anthropological evidence, had been calling, each to each other, to be gathered in one place, for what esoteric or eldritch purpose X couldn't fathom. Adrift in human time, these amnesiac artifacts had commissioned X as their human agent in this attempt at self-recollection. X, the human with the freakish talent, the penetrative gaze, enlisted as their emissary and agent among themselves. Their knight and champion, their uniter. Yet like any such foot soldier, X only knew what he needed to do his work, was quarantined in his cell of one from the longer agendas, the greater outcomes. Was likely destined to be discarded, too, by those he'd served, in whose company he'd never belong.

For if there was one thing X had come to understand it was that for everything his gaze revealed to him there remained limits, severe and unforgiving ones, to what the objects could tell him, and to how completely he could identify himself with and inside them. No matter that they were strange and he was strange. No matter that he loved them. To confuse himself with what he saw into, and with what he saw inside when he looked inside, was a mistake. The artifacts were not mirrors for him. That vanity had fled X, years before.

. . . .

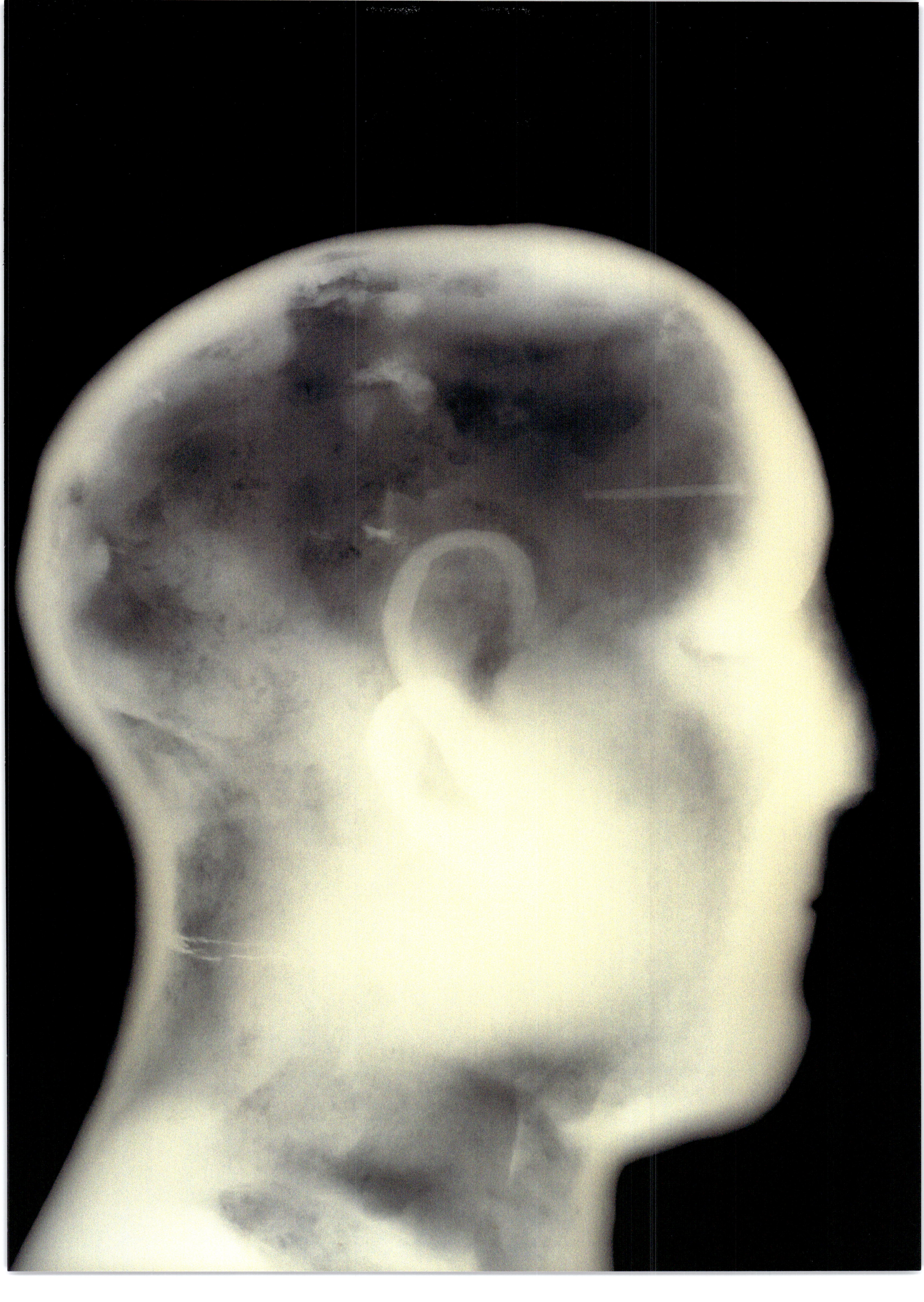

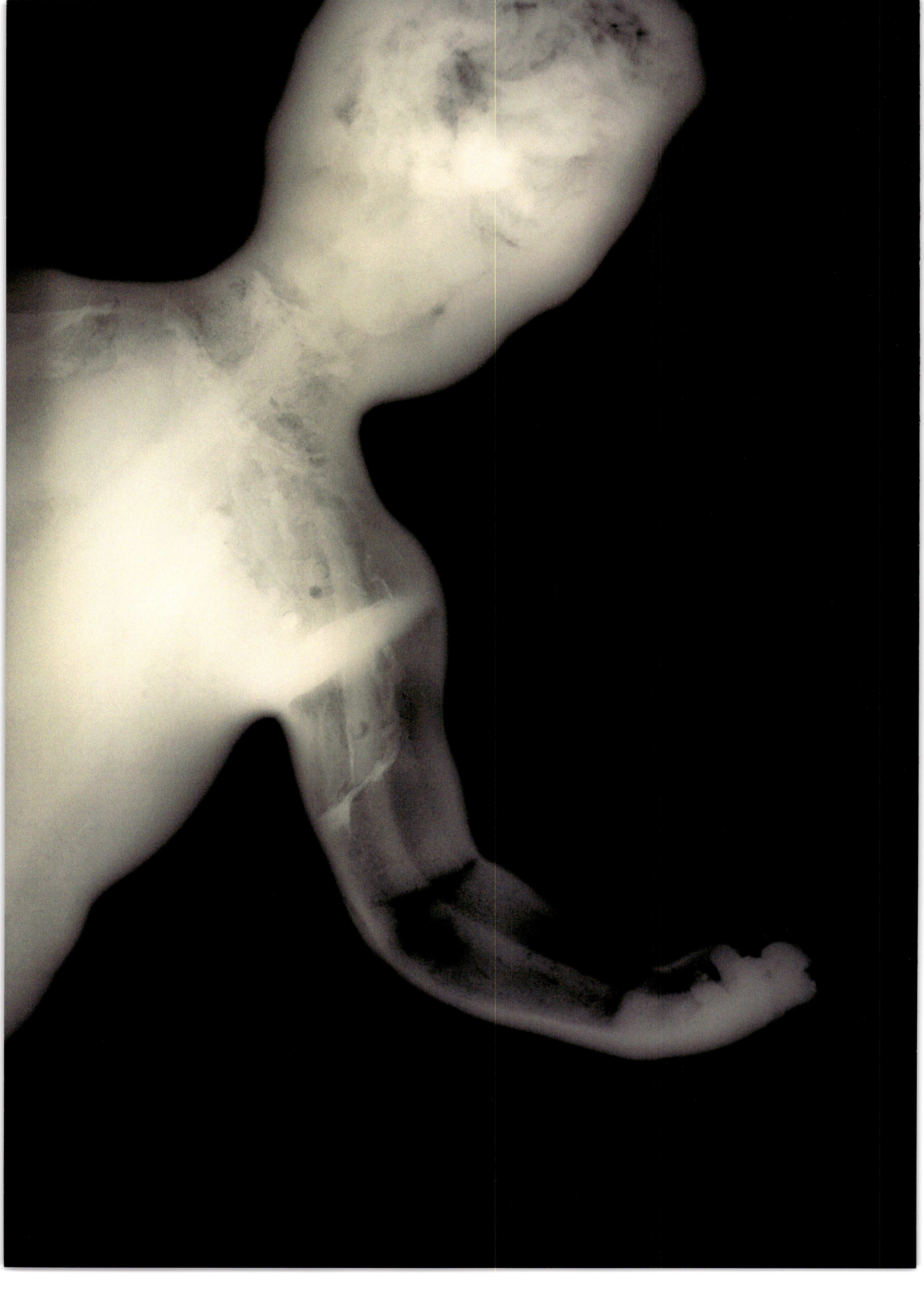

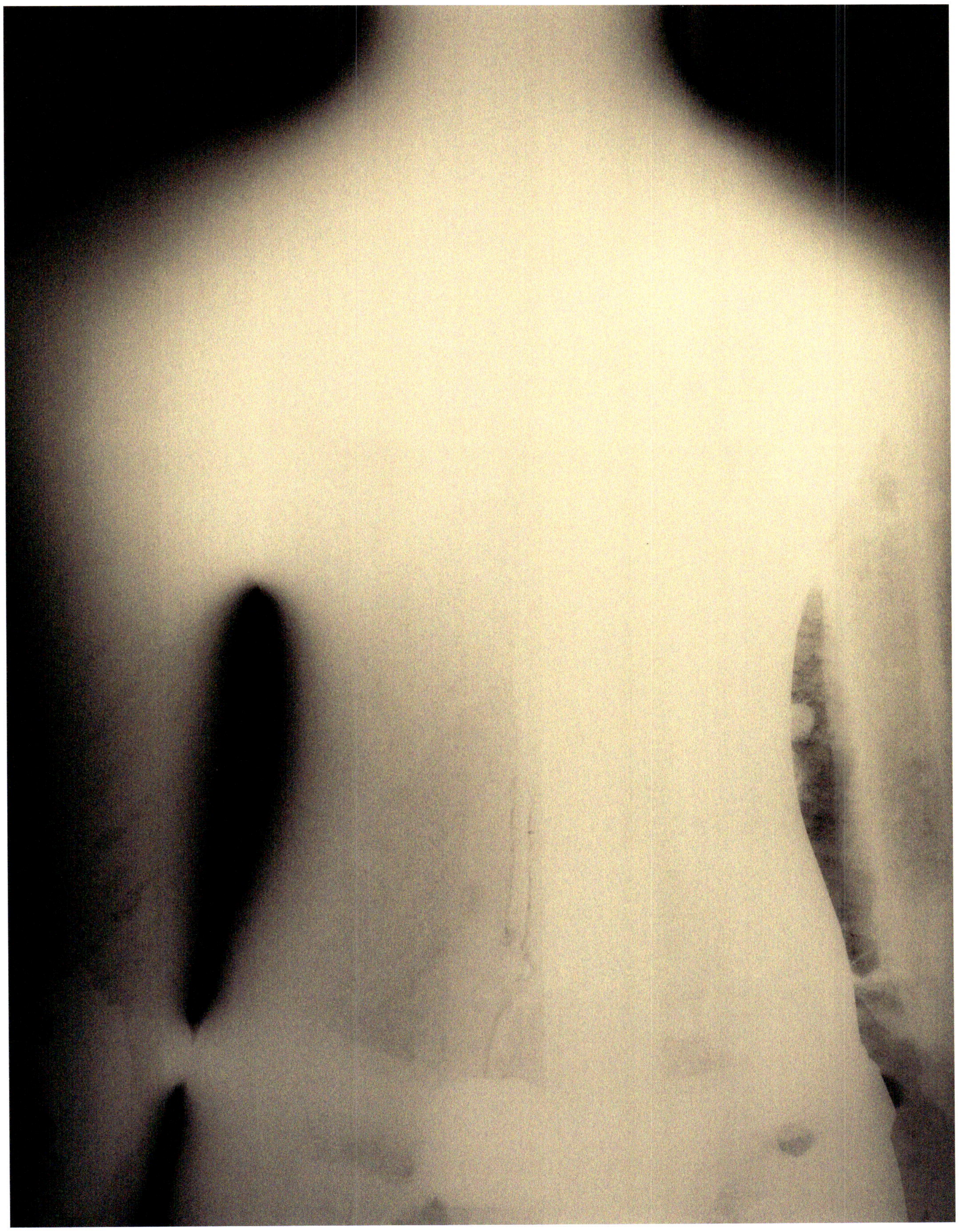

GM13 / AB14

GM25

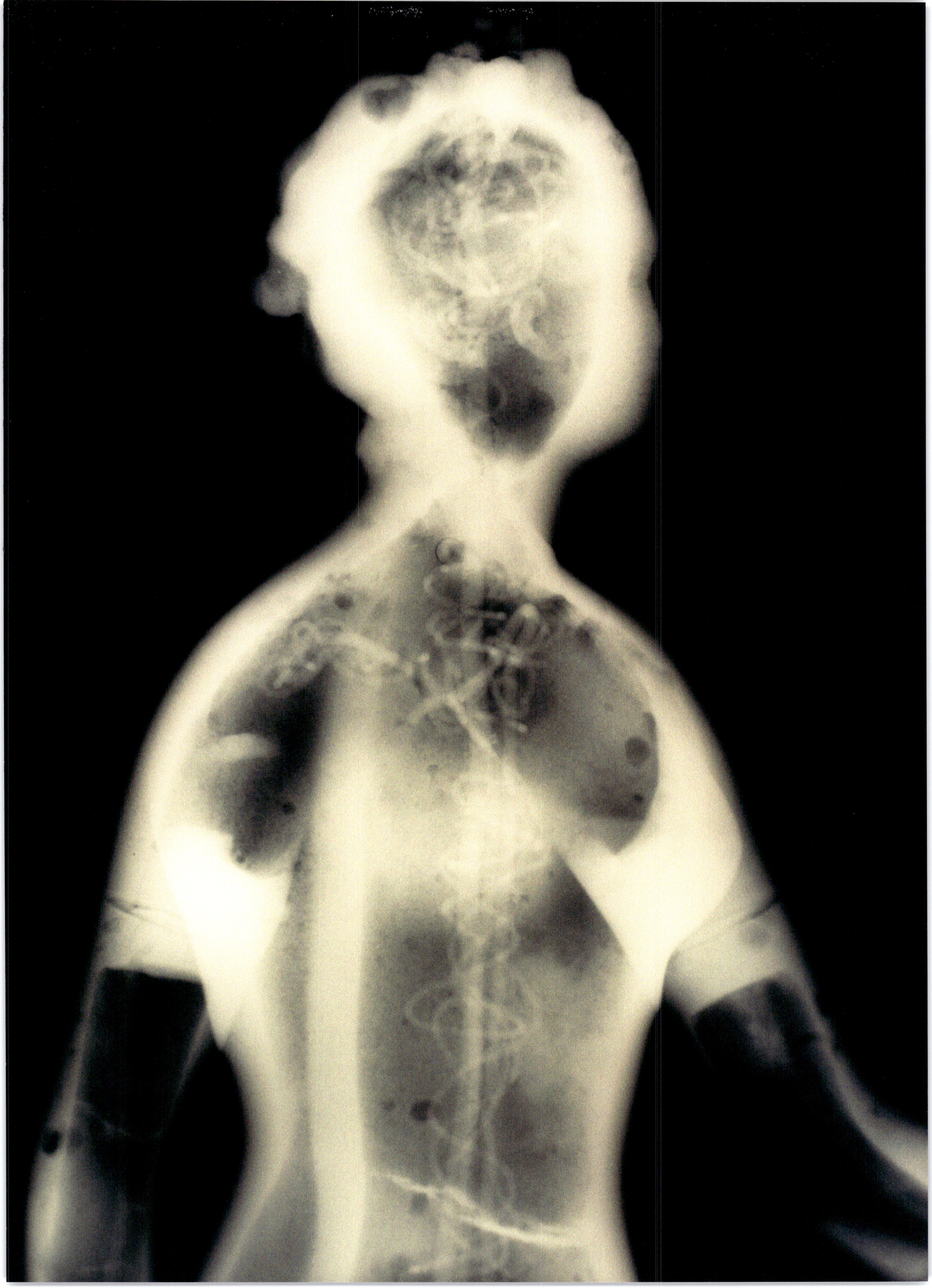

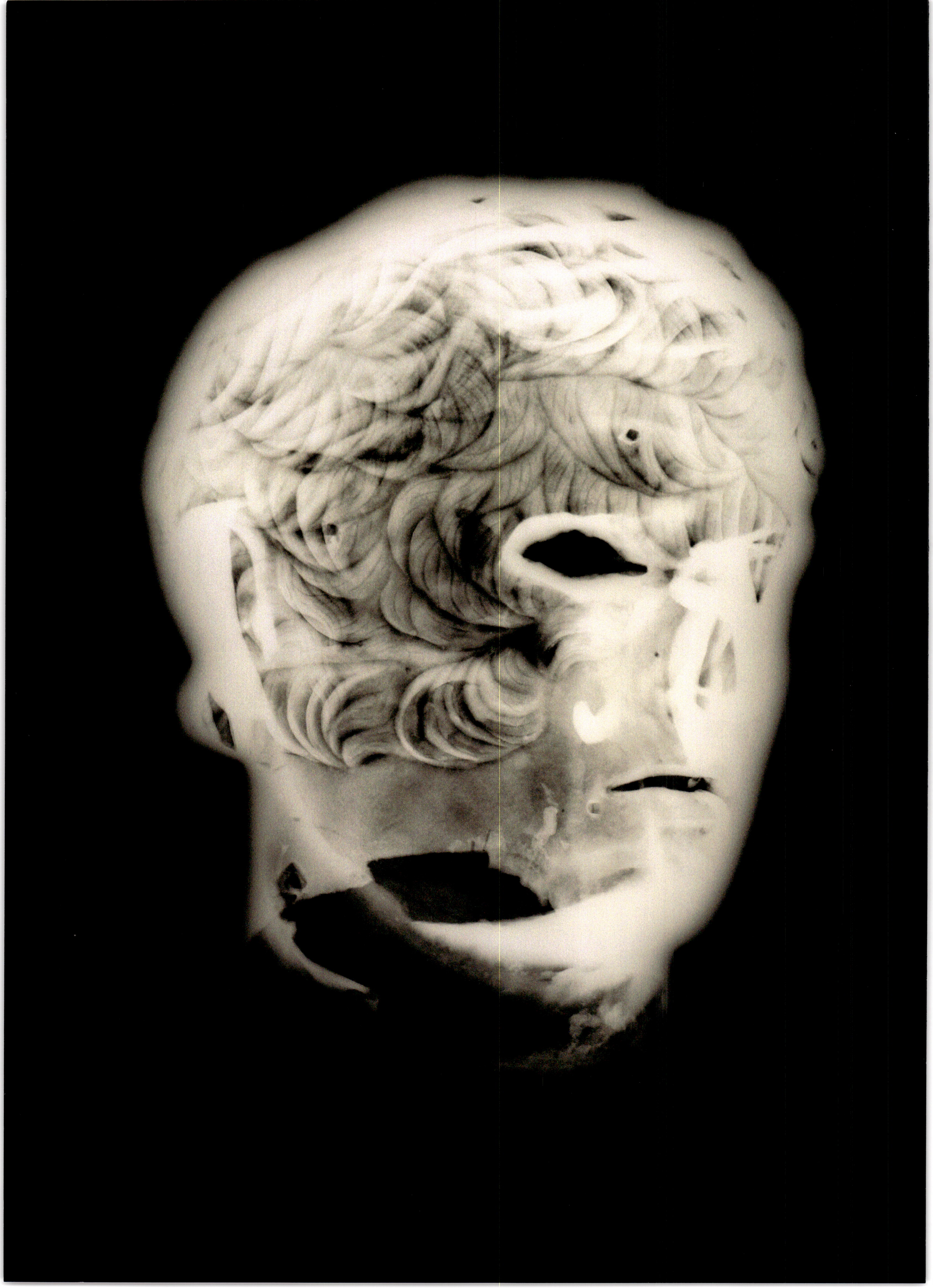

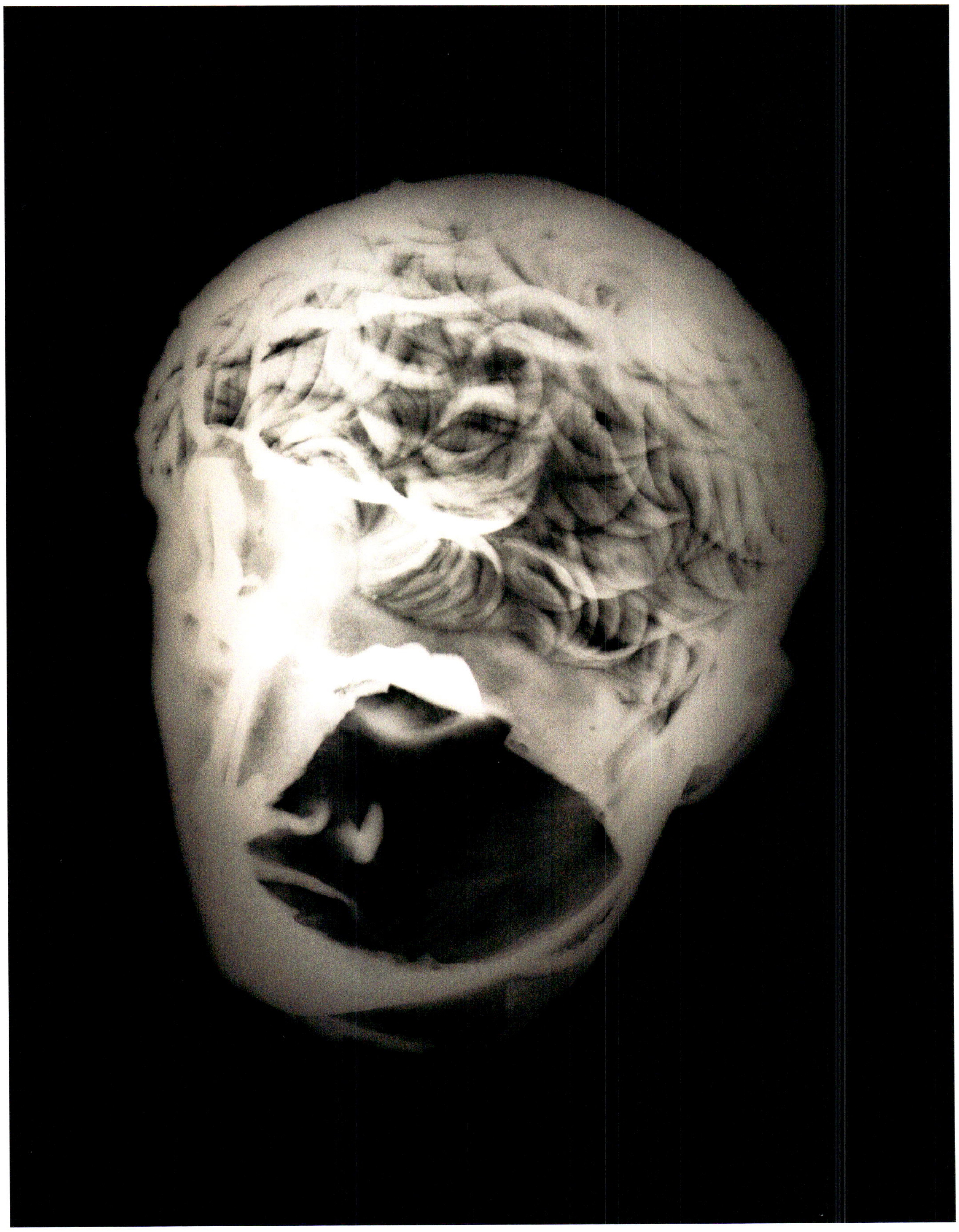

GM2 / GM17

For the uncertainties surrounding his commission, X recalled as if it was yesterday the onset of his gift. First, of course, came the wishing. The wishes he wished hardly made him special. In fact, they made him as ordinary as any child of his century, perhaps of any child ever born, though the words and notions a child would have used to express this wishing in the time before the century of photography and radiology and the H-bomb would have been entirely different. Impossible for X to know, since he was wholly formed by those things, a generation's metaphorical framework. I want to see my bones inside my skin. I want to see through her clothes. Her dress, her underwear. I want to see through the walls and floors of my bedroom when I'm exiled in the dark and the grownups go on with their lives. I want to know what makes the television work. I want to know what makes the kitten work. I want to tease apart a stone or a cake of soap like it was a clock and see how the pieces fit together, if they still do.

The world even tantalized a boy like X with its suggestion that such things were readily available. He'd seen the ads, alongside those for ten-foot submarines and sea monkeys, promising eyeglasses that would deliver the vision he craved, though he knew to be skeptical of that age-blurred huckster's iconography. He might as well try not only to raise sea monkeys, but to fall in love with the pretty one. There were movies too, Ray Milland or Superman burning holes in bank vaults and lightly scandalized females. Teams of secret agents or rebellious vagrants who donned special Ray-Bans that unmade the gauze of lies which coated the world, stripped away conspiracy to reveal the evil alien invaders who cavorted in the midst of oblivious humans. X's delusory hunger for eyes that unmasked the secret life might evidently be so universal an appetite as to lose its sacred appeal altogether. He didn't want to be just another fool, another actor, glaring and arching his eyebrow as though spotting something he couldn't, in on a secret that didn't exist. Yet just at the moment he felt this heretical disinterest, everything changed. X's father took him to the museum.

. . . .

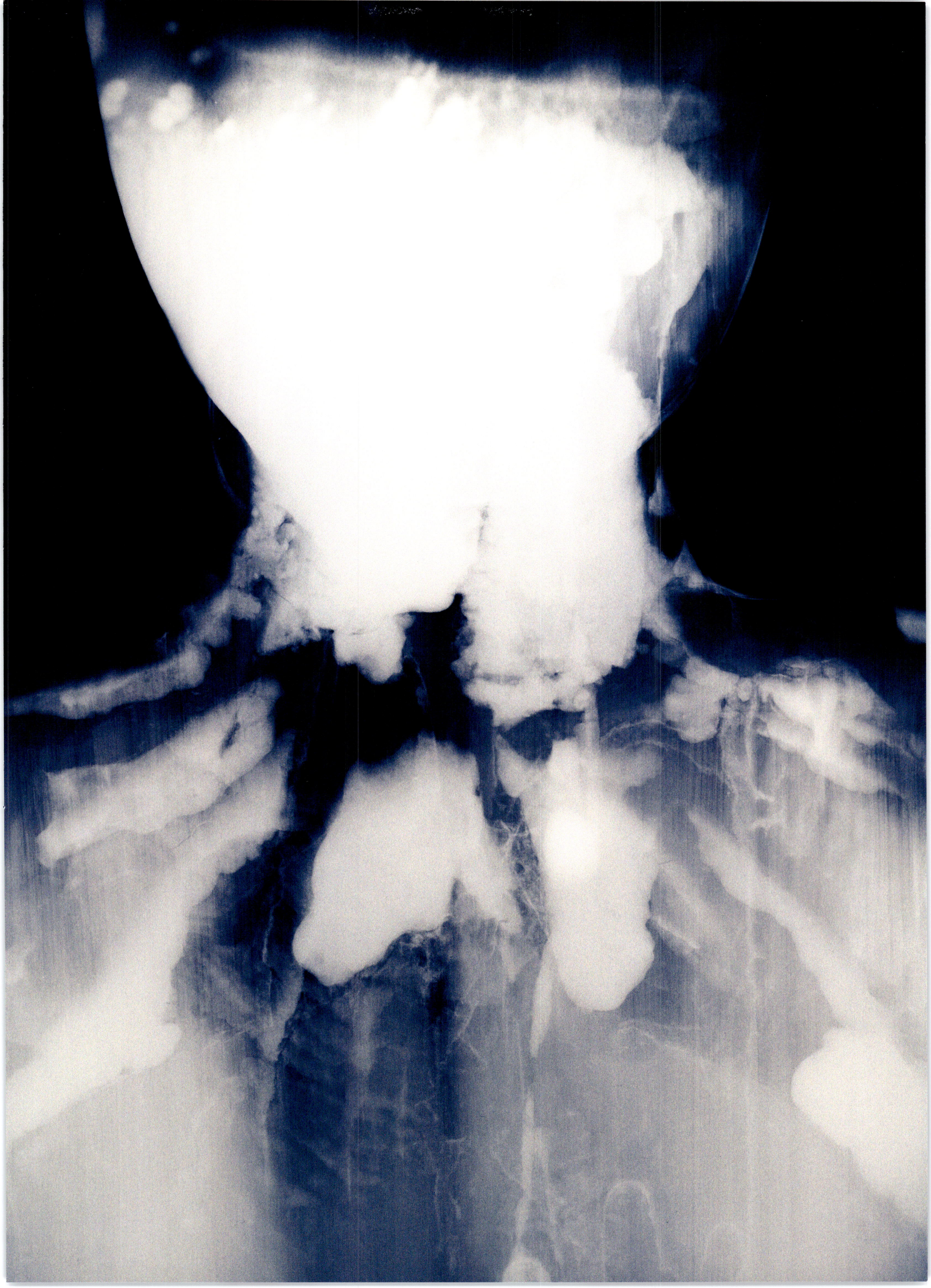

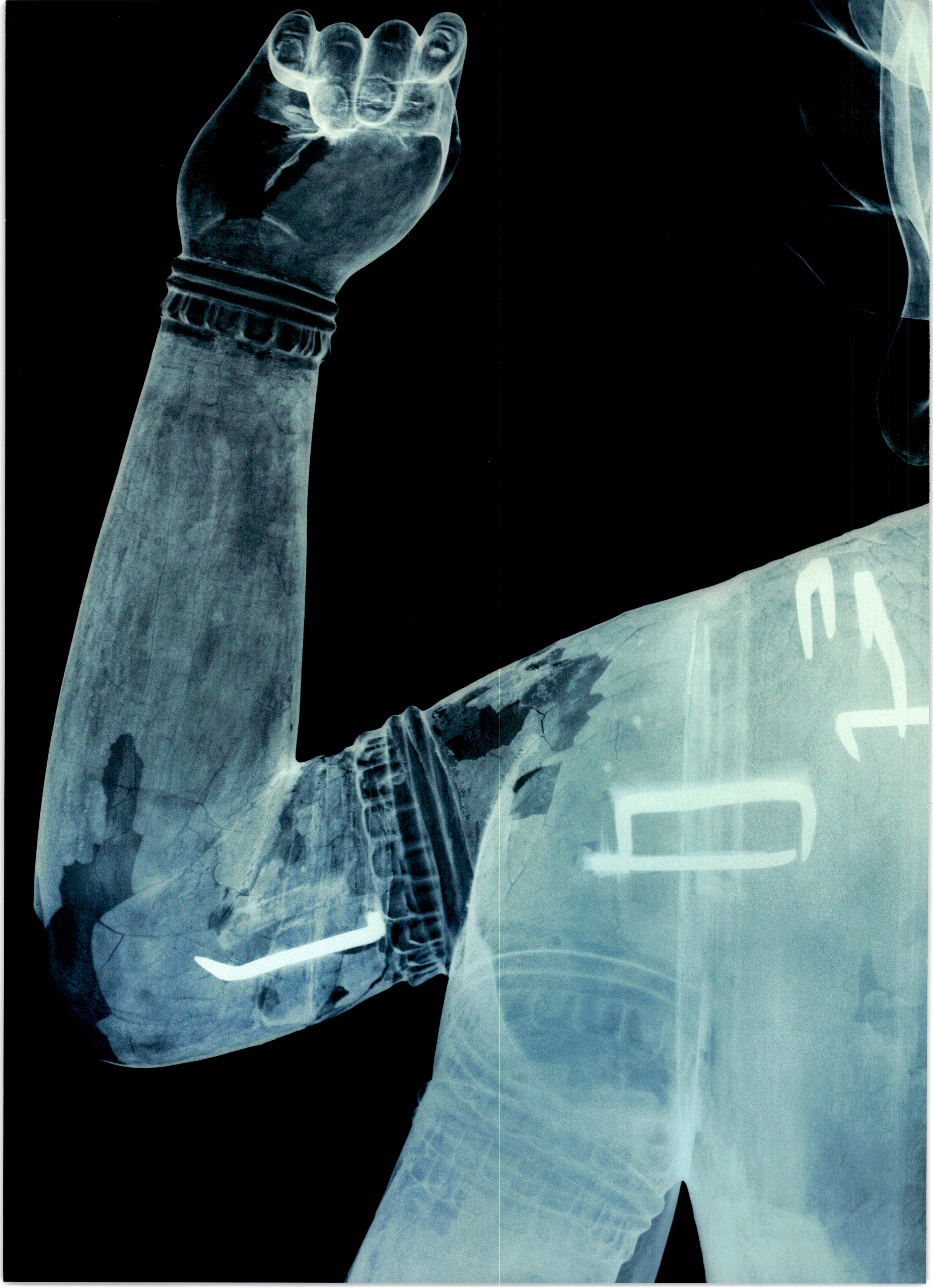

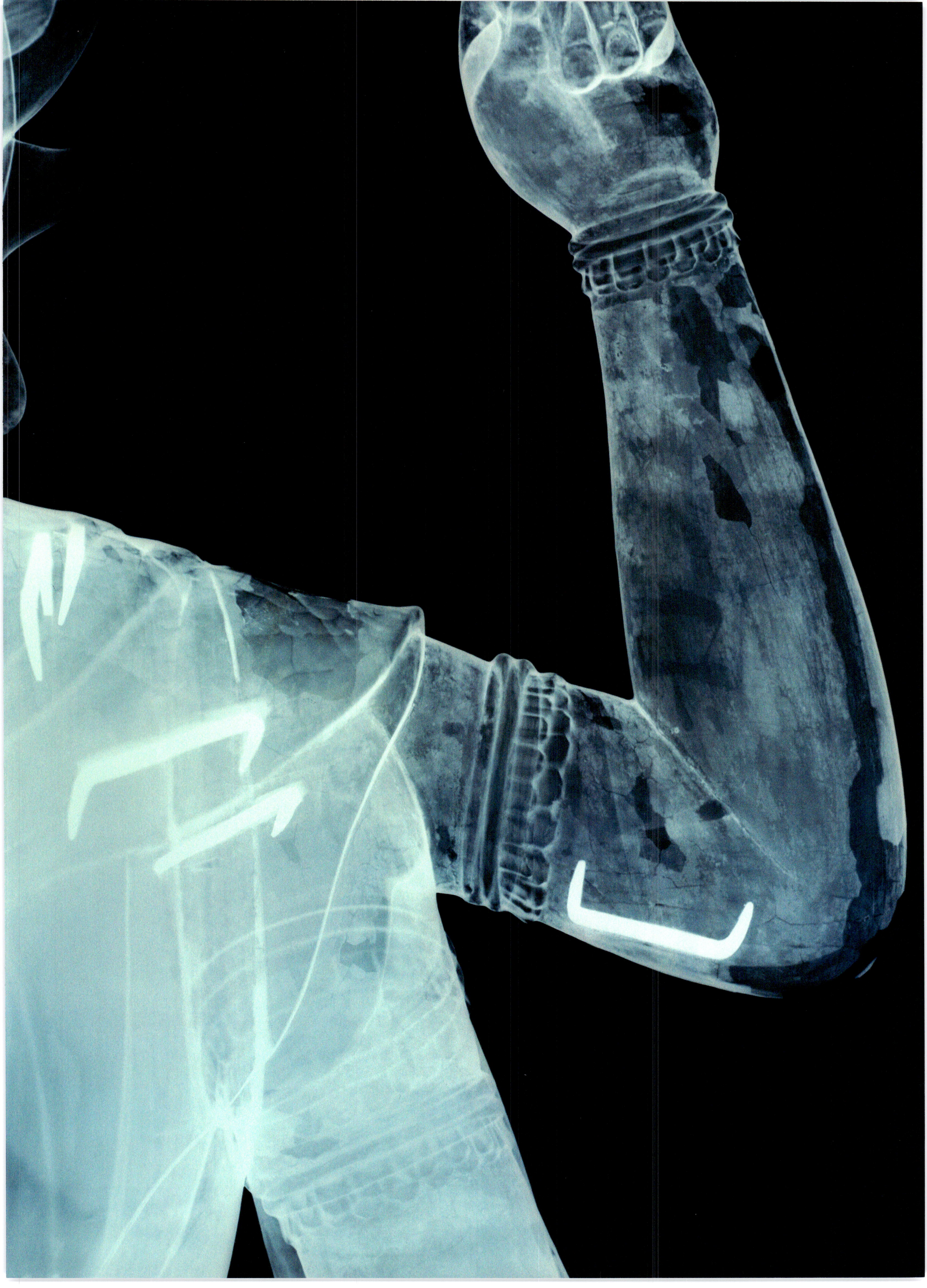

The building itself was a kind of conundrum of monumentality, an ark you entered through a porthole at the bottom, an Athens you invaded by means of the Trojan Horse of your self. X felt like an electron sliding into the gate of a collider, about to be expanded or imploded by some unfathomable science. Yet his fascination with these surroundings melted away instantly upon contact with the first of the artifacts susceptible to his gaze, the bronze statue that taught him how to see, or taught him that he already knew how but had not until this point encountered anything worth seeing. At this encounter, X's eyes graduated.

It was a horse. The sheeny bronze collapsed into a series of corridors and doorways, portals to an interior that was also an autobiography of the horse's making and its makers, a constellation of tiny views, map orientations and carbon signatures of its movement through the centuries and from its point of origin, a shadow of some forge on a hill, a rain-streaked granite pedestal, a half-eaten feast on a table under which some irresponsible emissary snored out a red wine hangover. The horse hummed with a wordless sense of its fate and journey that outshone the scatterbrained buzz of everyday circumstances. Soon enough it had companions, too. A chalice fingerprinted with interior architecture, a dense musical notation or mathematics X recognized without discovering its name. Two women made of marble, interiors laced with human trance leavings, with dream script.

Not every object in the museum spoke to X's new eyes; whether the failing was X's or that of the artifacts he would never be sure. Perhaps there were other seers free of some flaw or doubt who'd have found the whole museum, the whole world, lucid in this manner. As it was, dozens desublimated themselves into his gaze, while hundreds of others remained inert, at least for the time being. Possibly this was simply a matter of tolerances, a limit on what he could take. The majority simply slid off into the gray haze that encompassed the museum corridors and floors, the other patrons, his father.

The new capacity followed him out of the museum. Various items throbbed with secret vivacity, distracting him from various contexts to which X had been supposed to attend: a French teacher's onyx bracelet, the soda tap at the drugstore counter at Macy's where his grandmother liked to treat him to Saturday egg creams. Nothing with the potency of the museum's treasures, but how could they be expected to have attracted comparable resonances in so short a time? X felt a forgiveness toward the feebler emanations of the newer items. It was enough that they stirred even slightly, their rustling presence suggesting they'd gained first inklings of their place in the historical pageant of objects.

. . . .

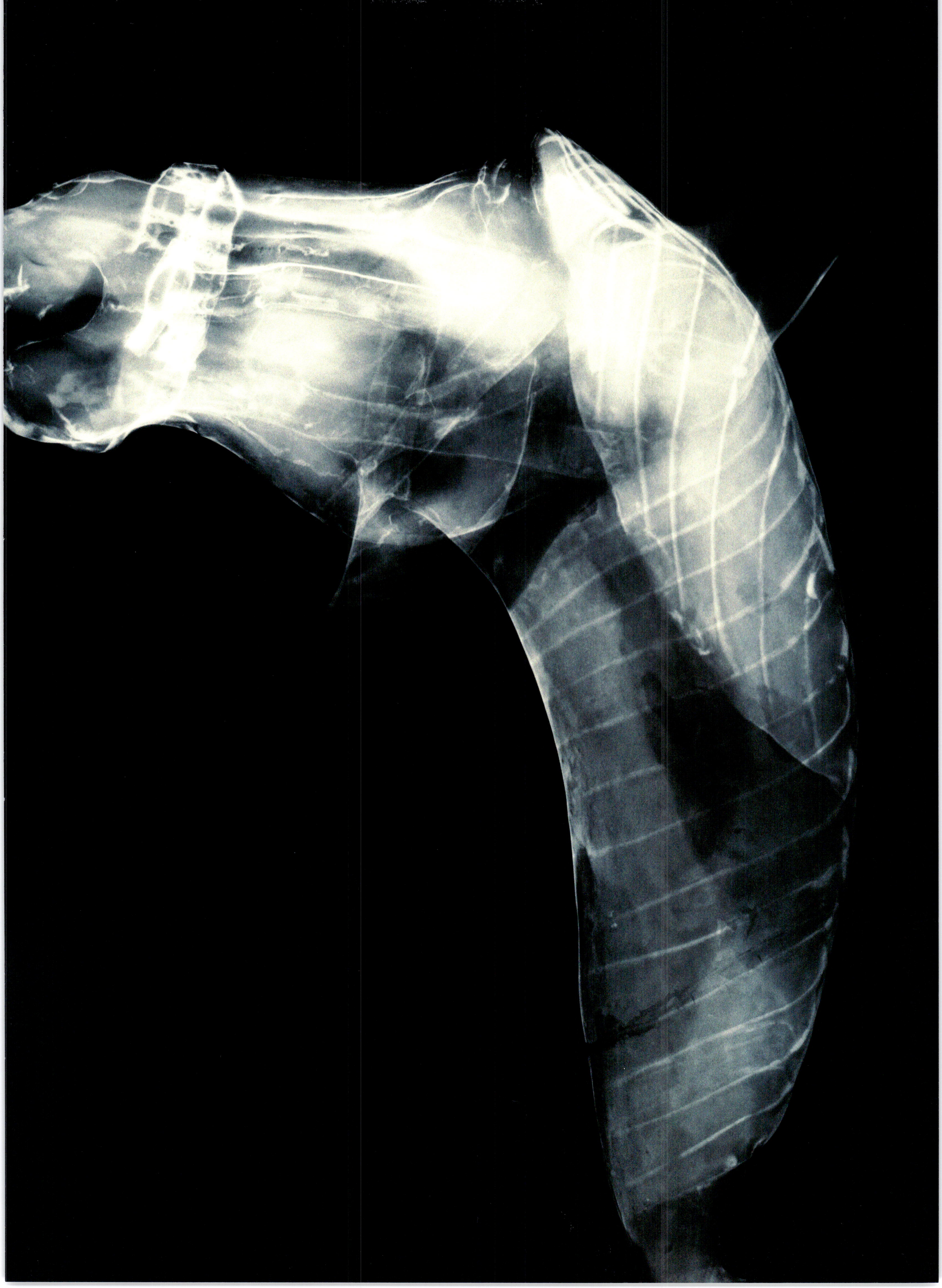

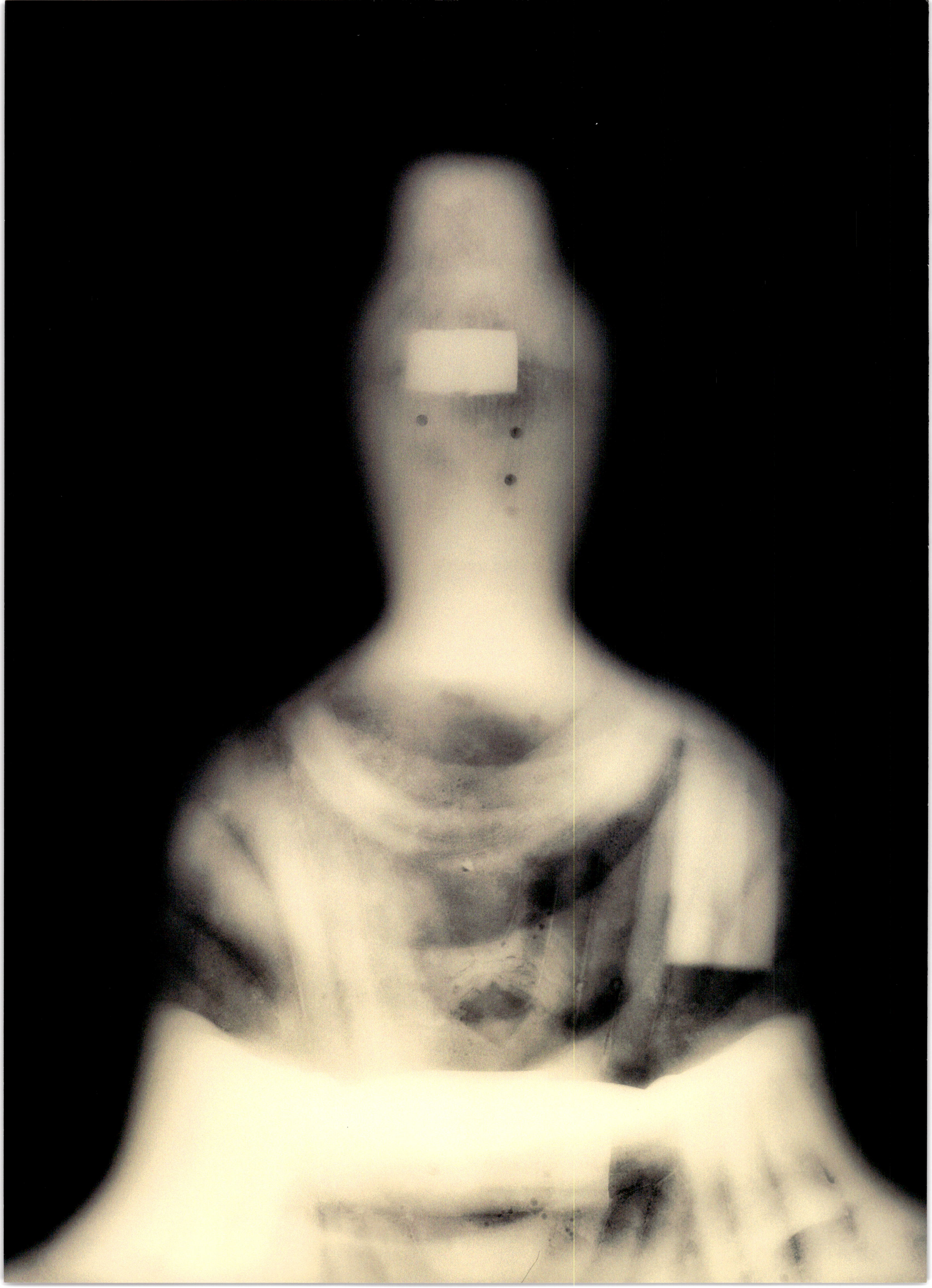

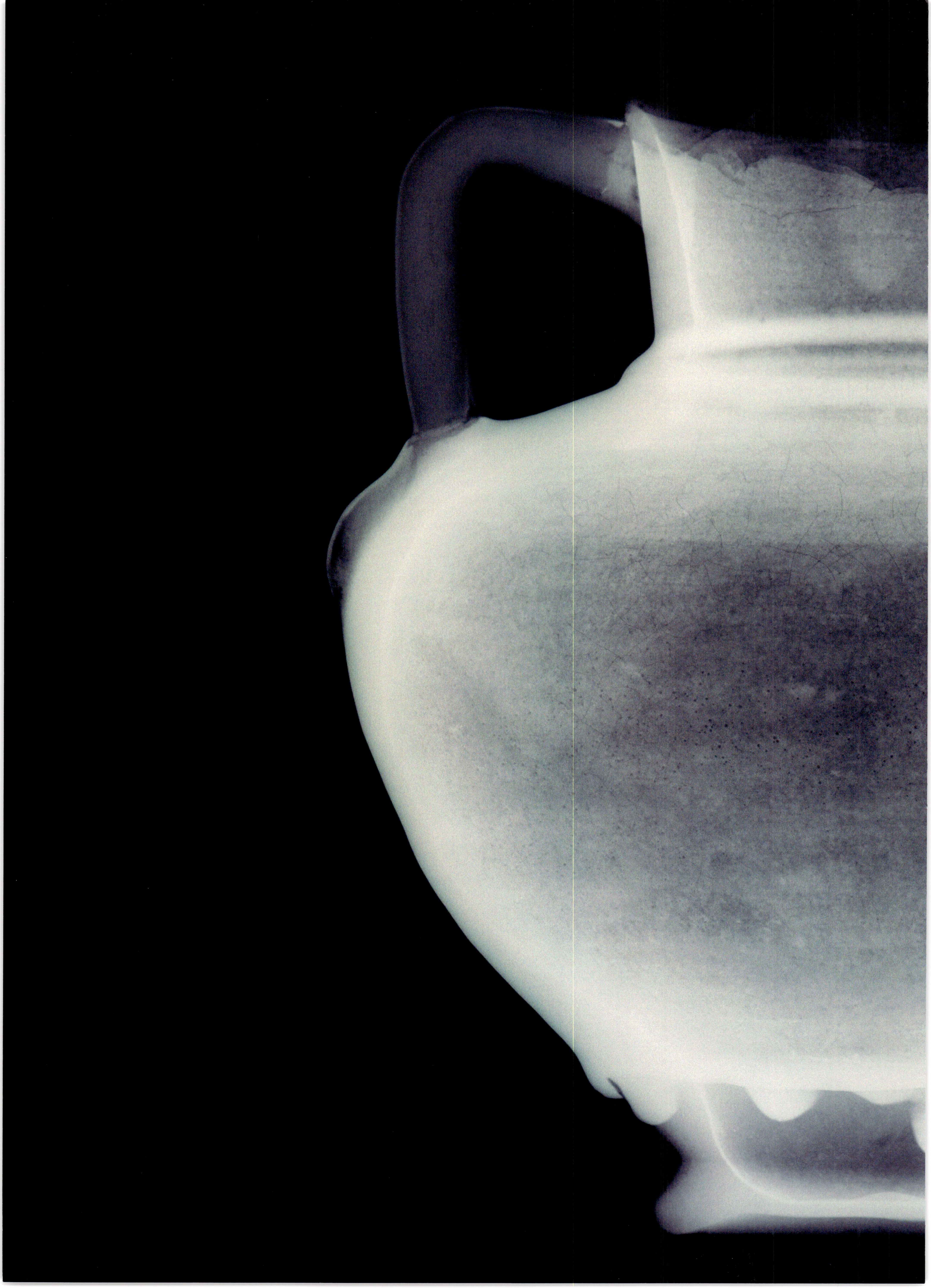

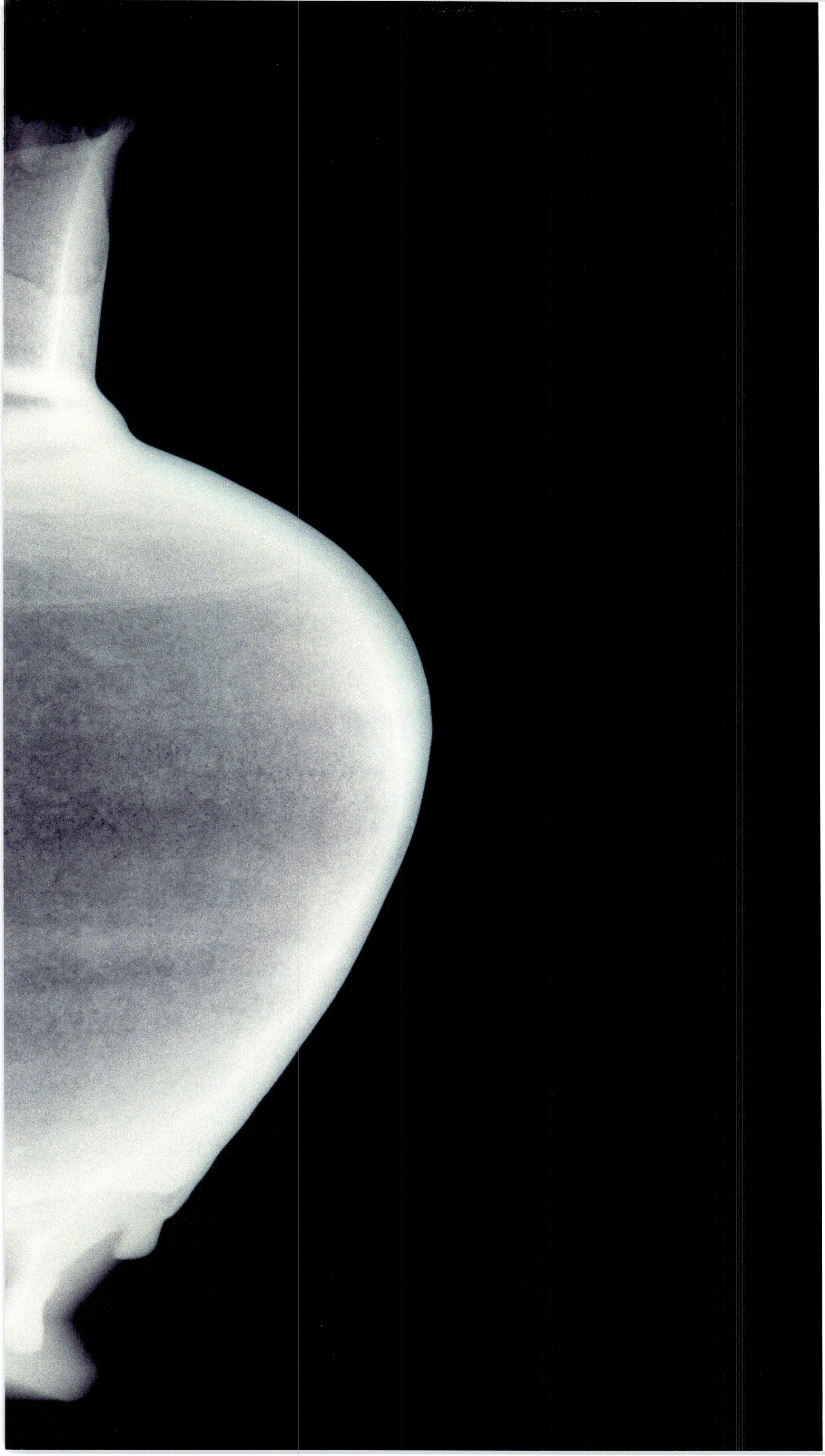

AV3 / GM12

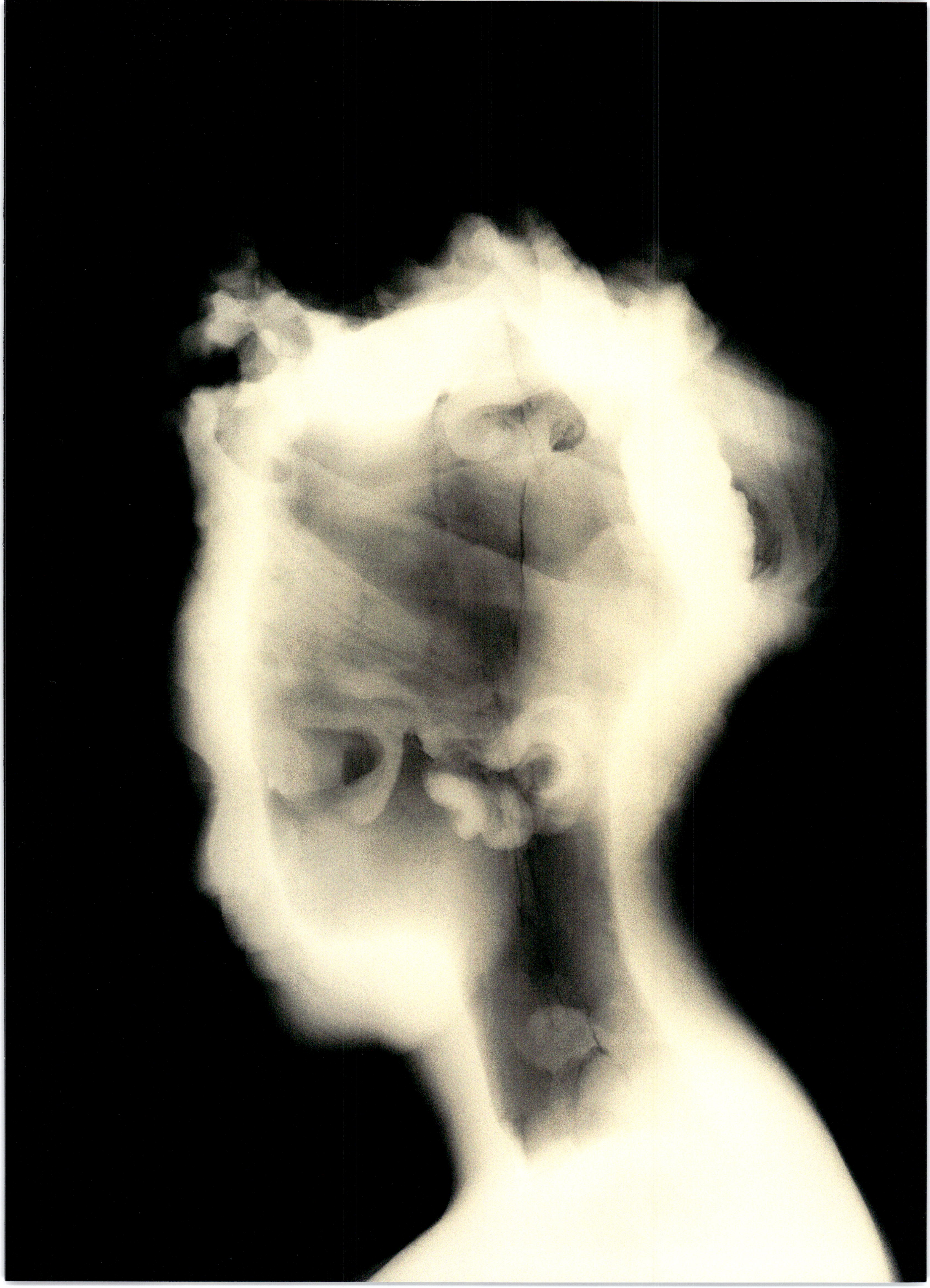

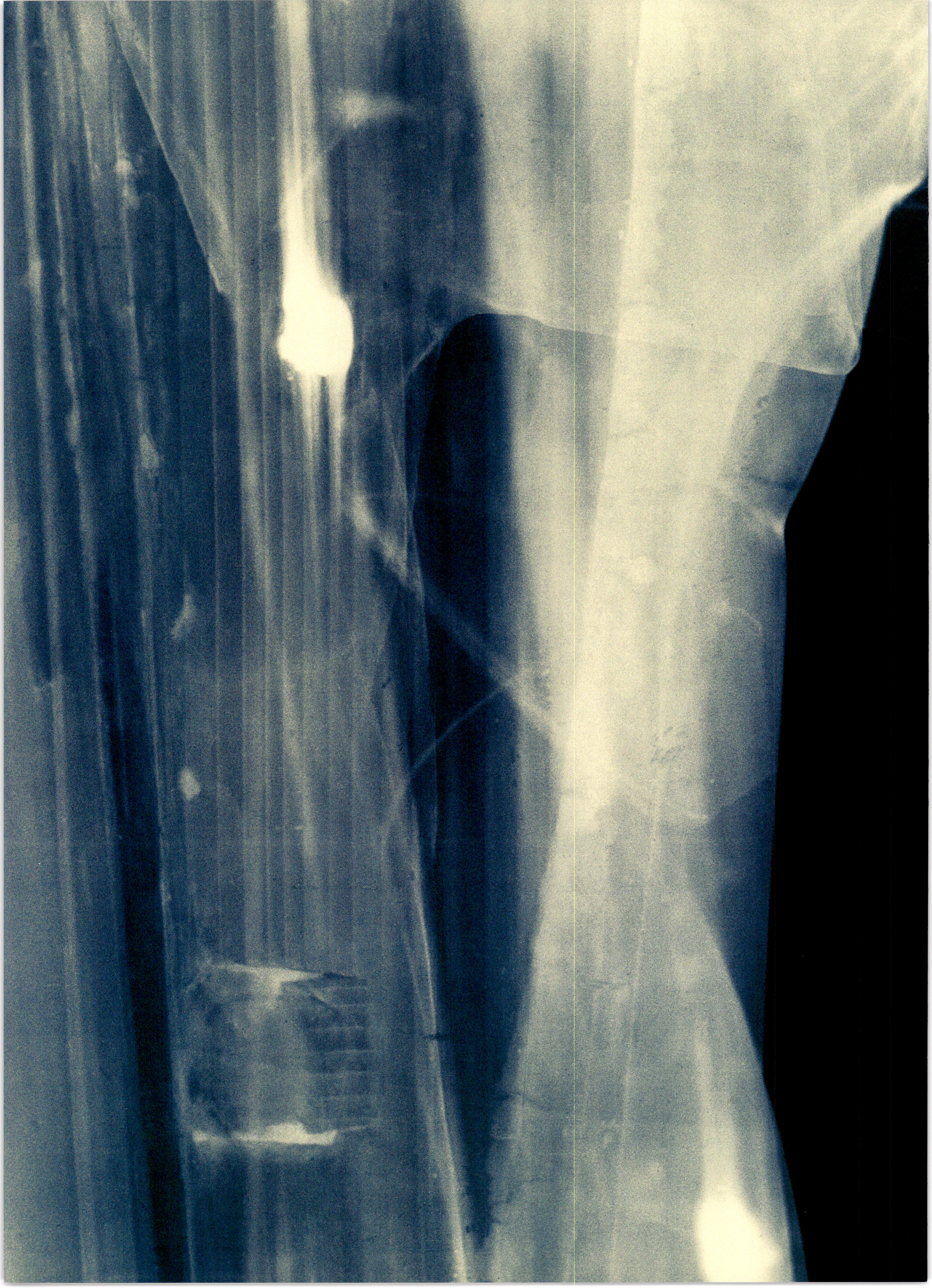

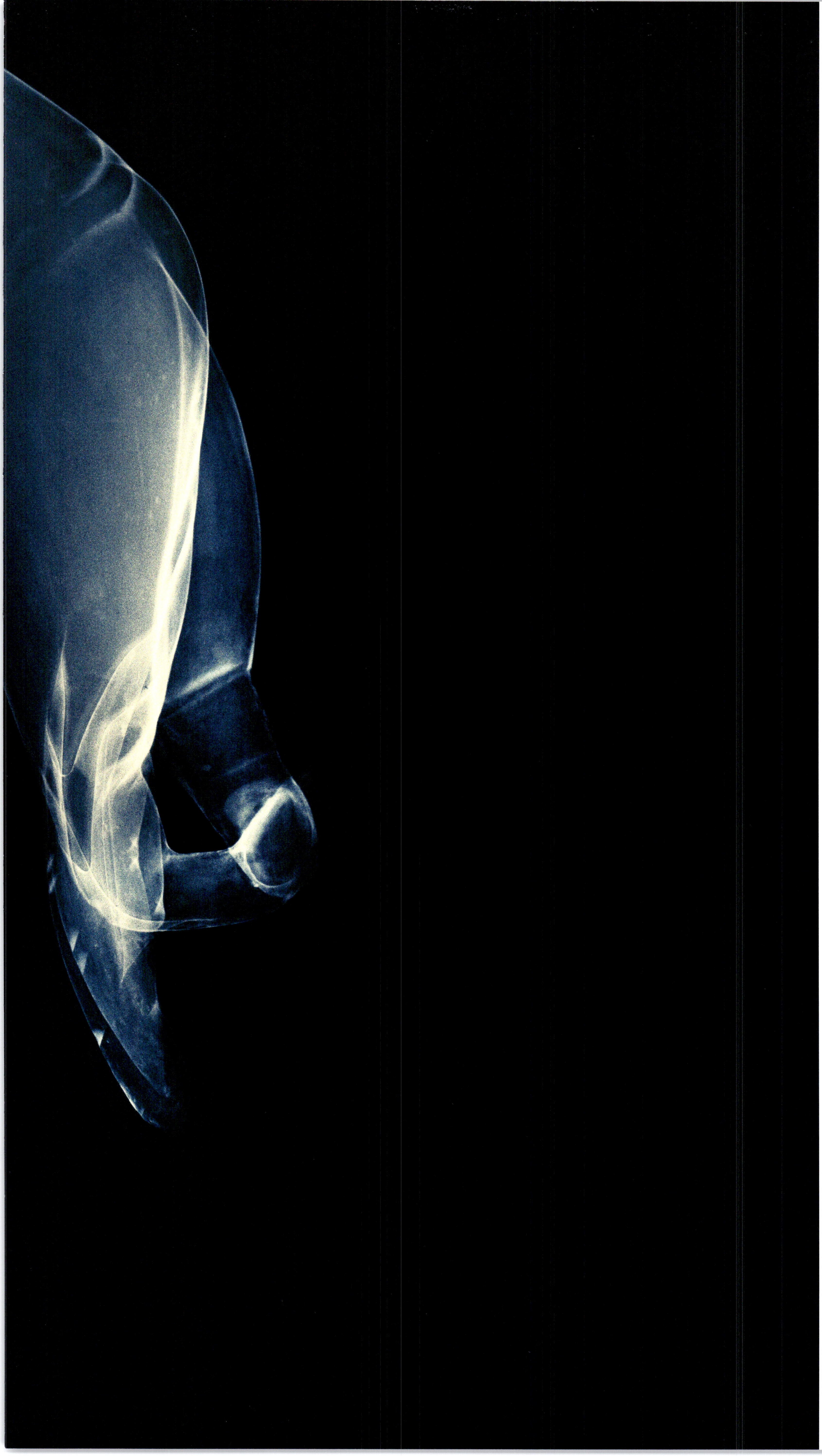

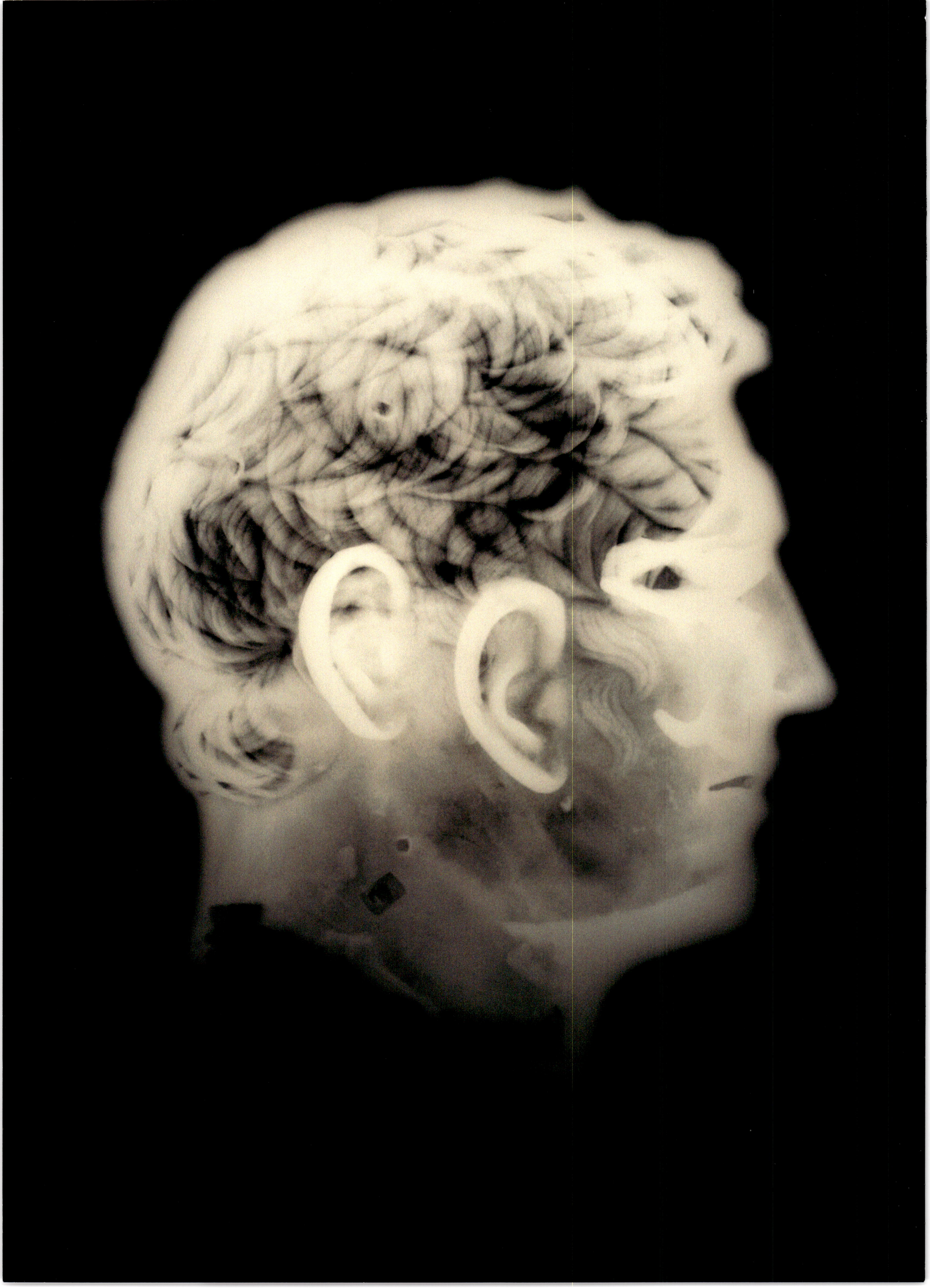

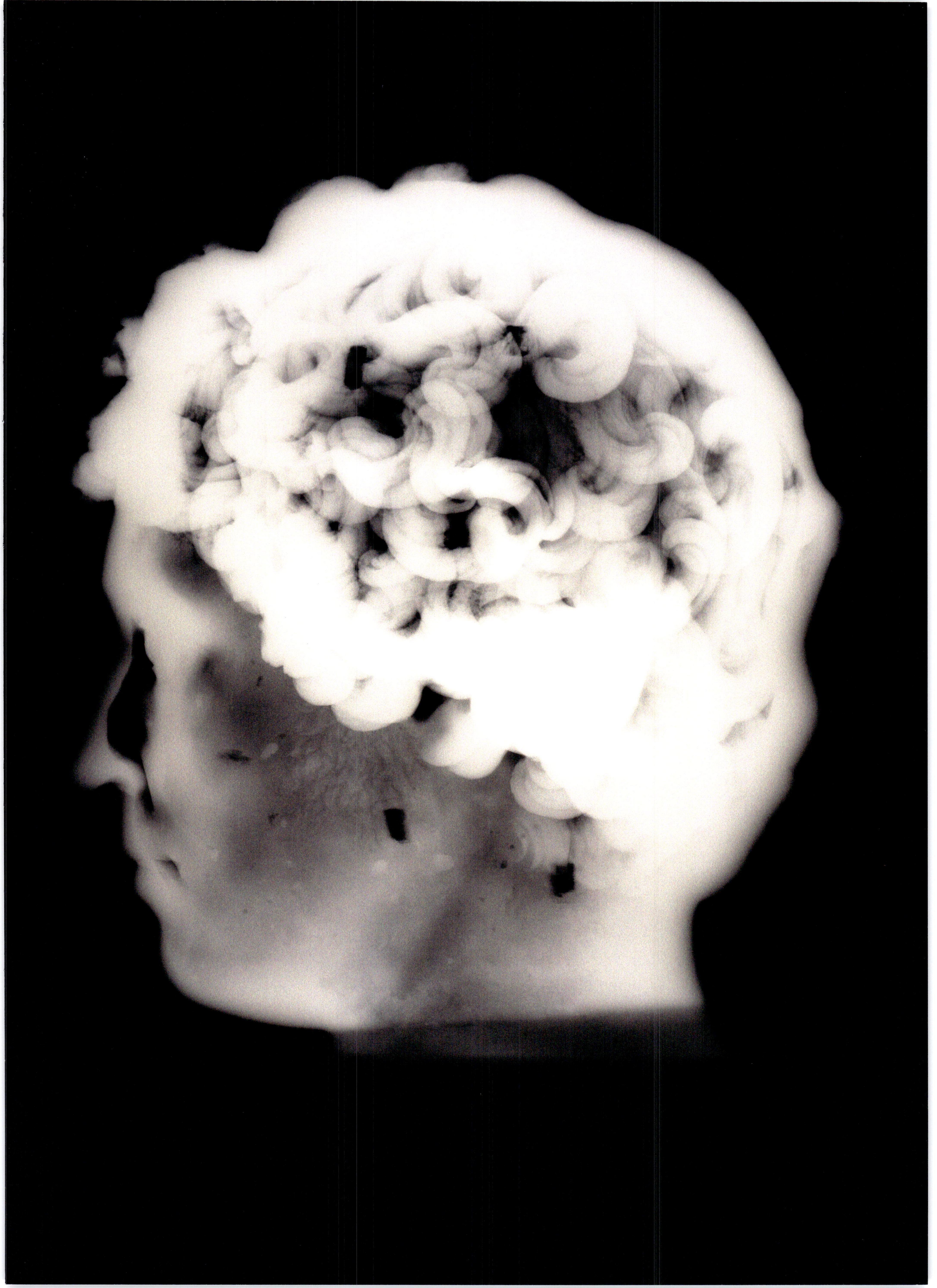

Never, though, was any person in any way enlivened or rendered translucent. The opposite instead. The clothing and skin with which human beings met the atmosphere might as well have been curtains of dull stone even to X's most lingering or hopeful glances. He grew more distant from his living fellows by default, as his relationship with the objects grew more enthralled and sensual. It was X's fate to live among the implicit traces of the dead, who were no more dead than the past was past. Yet though they weren't dead, the dead were somewhat more lonely than the living. It was precisely out of their isolation, and in order to slightly alleviate it, that they had located X with their siren call.

In his own loneliness, X yearned to know that others like him roamed the earth, sensitives distracted by the embedded potencies in select objects. A handful of cultural clues suggested he wasn't completely alone. The way Alfred Hitchcock lit a glass of milk and the very notion of the 'McGuffin'; Henry James constructing a fugue of a novel around the resonances locked inside that *Golden Bowl*; Stanley Kubrick's monolith. Sydney Greenstreet's rhapsodic covetousness of the Maltese Falcon, the lust in the fat man's eyes: even Humphrey Bogart, initially so skeptical that he dubs the Falcon 'the dingus', ends admitting it was *the stuff that dreams are made of*.

Yet by the time X was in the final term of a degree in art history—what else would do for him, he'd figured, but a career in the dark, floating his voice over luminous slides, trying to cause others to see what he saw?—he'd never met another seer, or at least never one who'd tipped their hand. X wondered if he gave any indications, himself, that another would detect. Probably not. He was given a jolt when a classmate, a brunette in bangs and funky cat glasses (a grown version of the type through whose dress his twelve-year-old self would most wish to have been able to see) offered up a disconcerting remark: "Keep in mind, you can't be deep without a surface!" He stared, wondering what she was getting at, but her smile was, as usual, opaque. The quip was almost reflexive, revealing less the more he considered it, yet, just as he might be about to dismiss it, seeming comprehensively relevant to his whole life. Was she being flirtatious? It seemed unlikely. Eventually he concluded from her lack of follow-up that the girl had meant nothing more than that he should attend to his personal grooming, which was admittedly atrocious. He took this to heart and reversed his slide towards dandruff, taped-glasses-frames, moth-eaten cardigans, and unboundaried facial hair. It was true after all that any object worth his excavating glance possessed a beautiful exterior to begin with. From that day forward X was more or less a dandy, his wardrobe and hygiene impeccable. Far from distinguishing him, this tended to make him disappear even further. Women now did flirt with X, but he never noticed. He was too busy flirting with the infinite. With one ambiguous remark the girl had transformed X from a geek at the mercy of his visionary gift to a cosmological spy.

. . . .

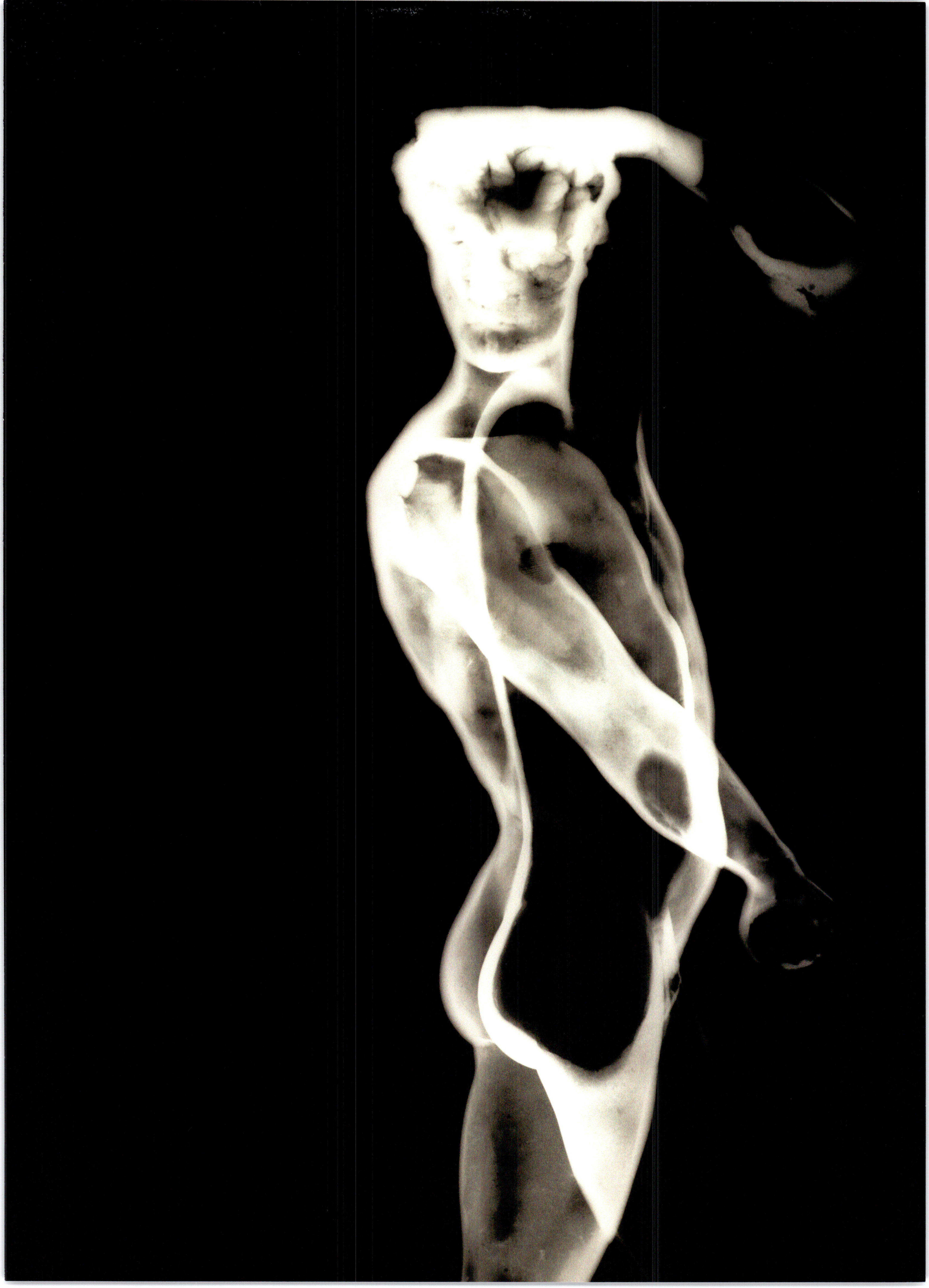

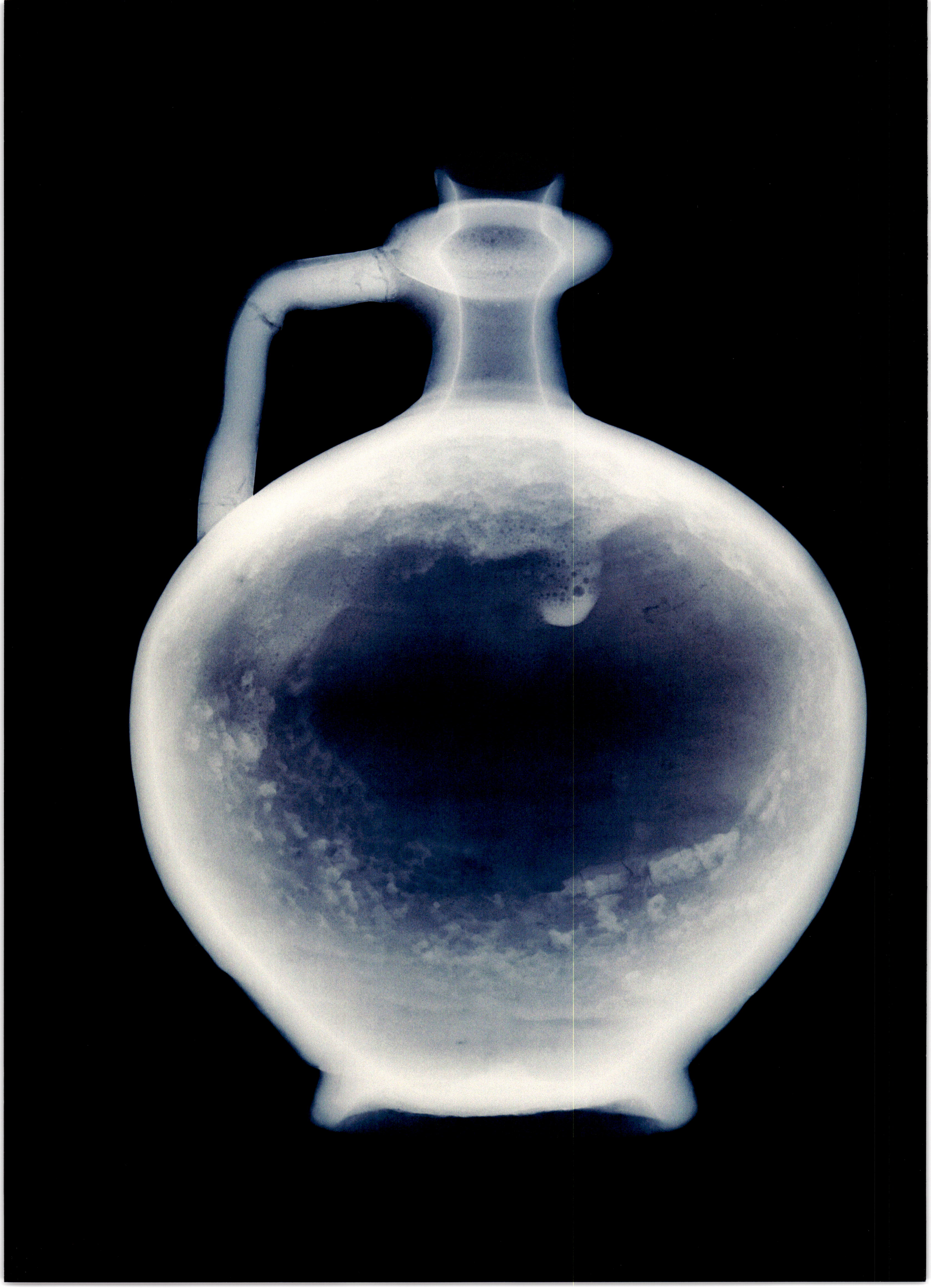

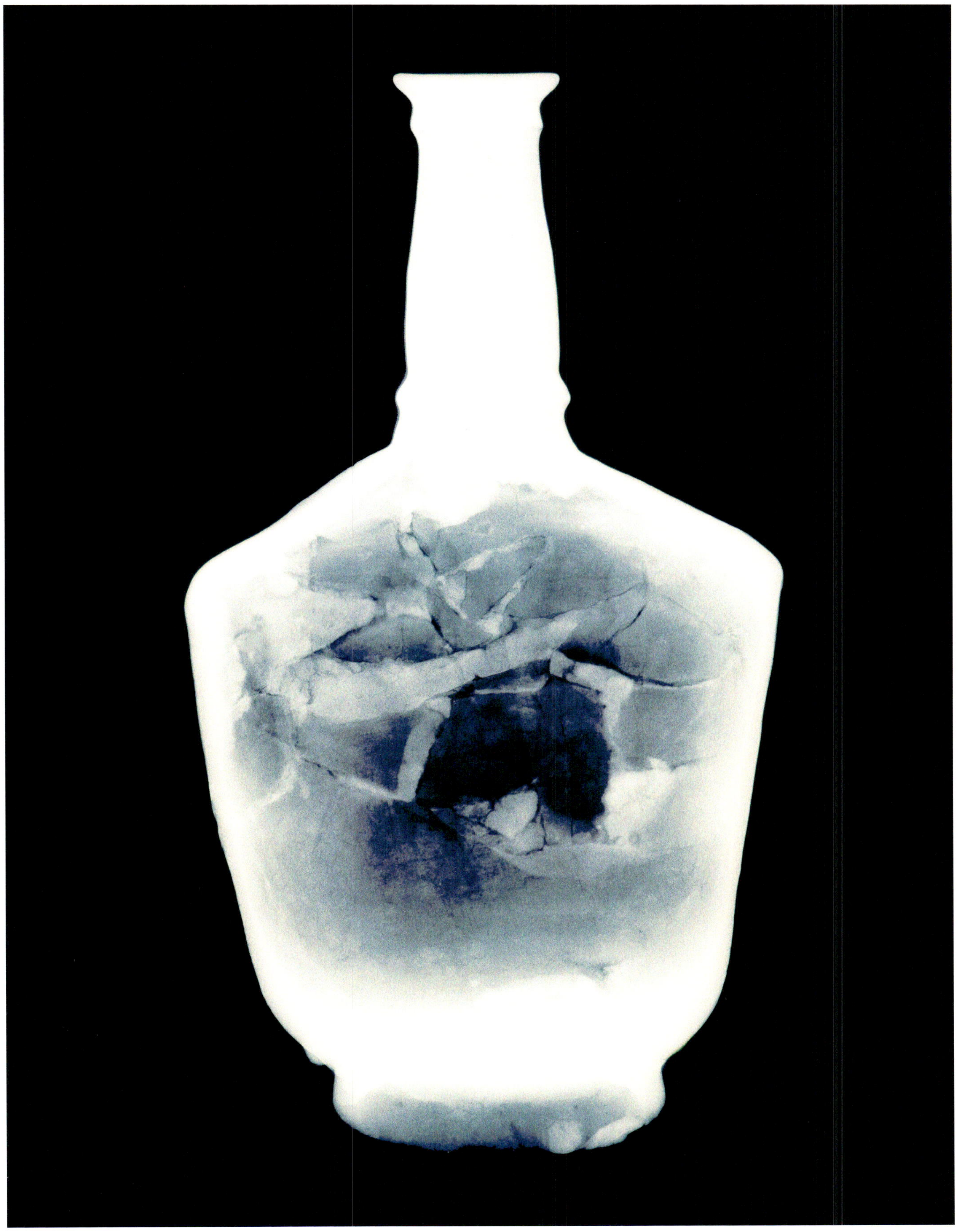

AV5 / AV7

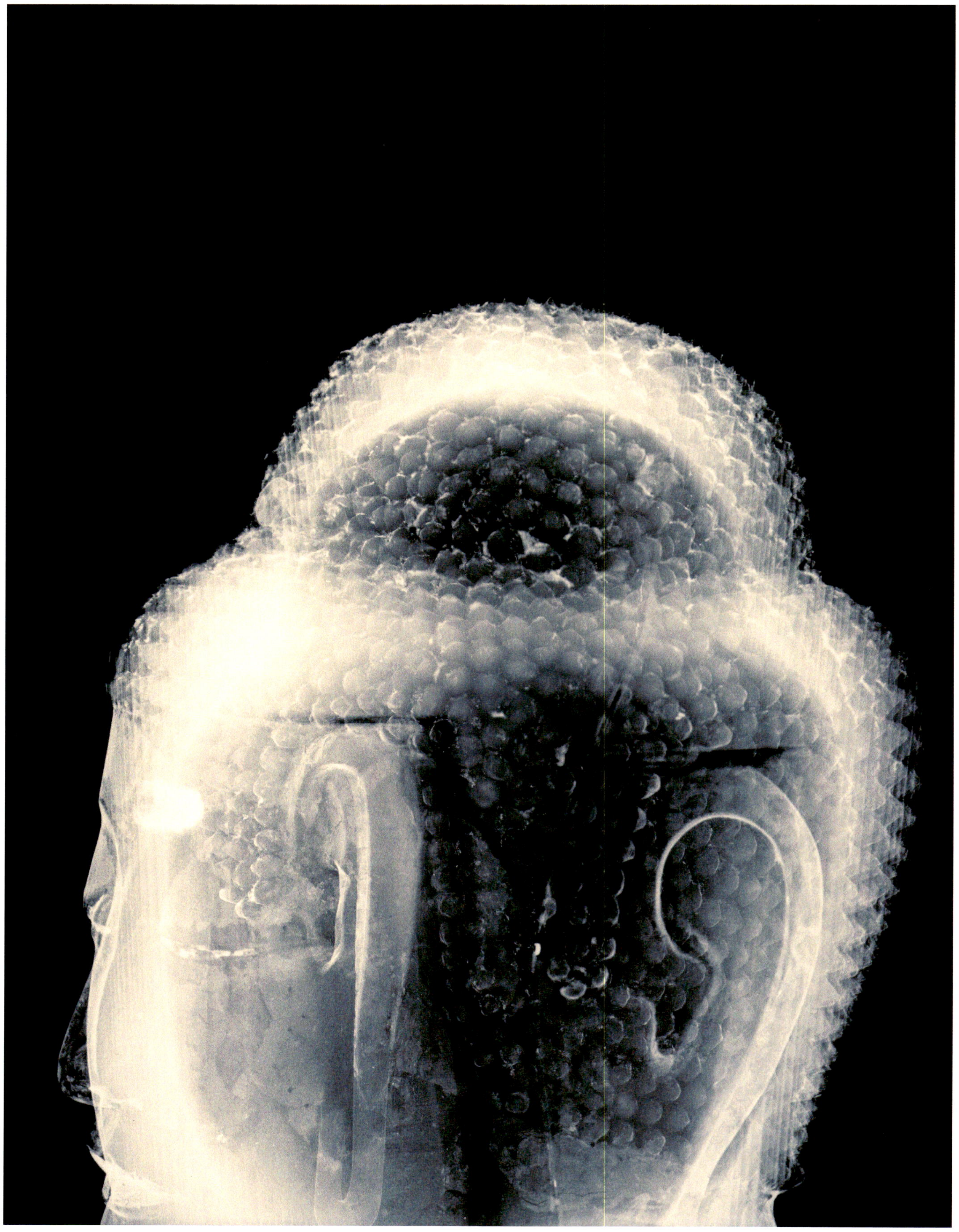

AB19 / AB13

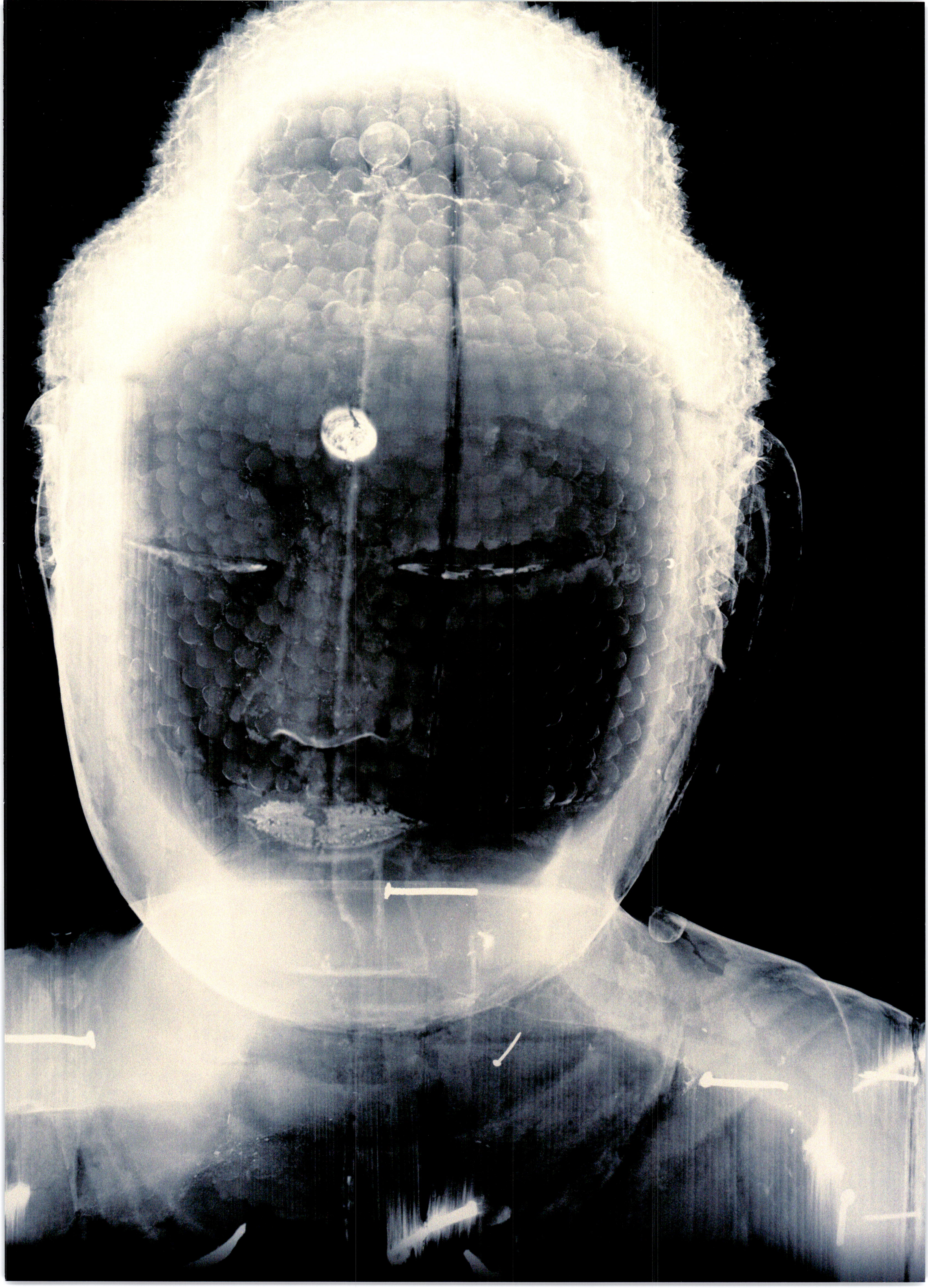

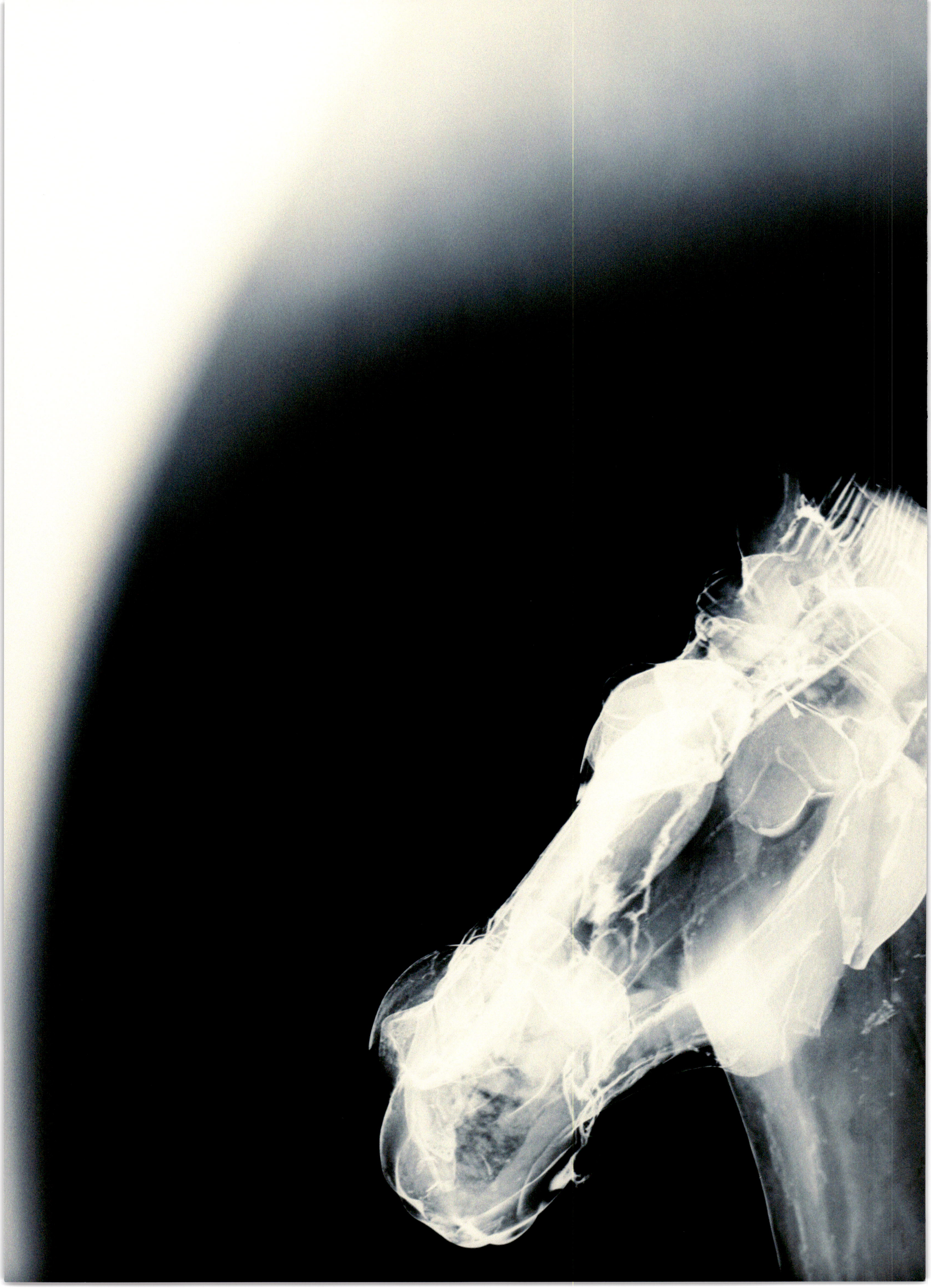

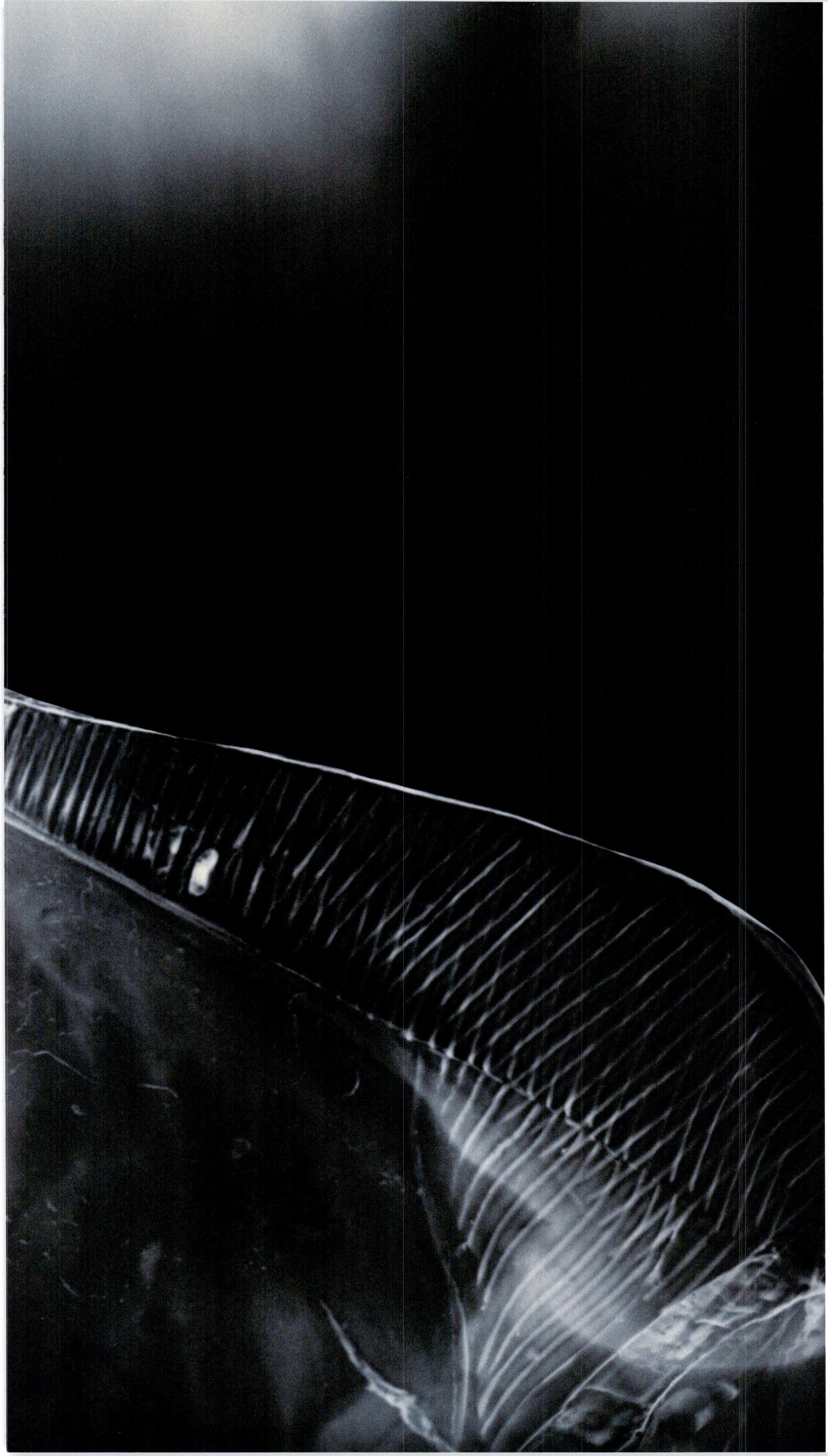

To see the insides and the outsides of things at once was to expose the falsehood of their division into a binary system. To diagnose the unstable systems underlying the apparently placid surfaces of the stuff that decorated the universe was not to reveal it as unsupportable or sick, but the opposite: the more you saw the more you knew that matter was tenacious and heartfelt, committed to its own presence on Earth, precisely for the specific and often homely circumstance of its coming into being, and for the tenuous nature of its apparent permanence. X was no doctor or mechanic, and his vision consisted of no procedure. He hadn't enticed the horse in the museum to drink barium, after all. Nothing needed healing, only to be abided with, as he'd begun abiding with the horse in the first place. Time's elegant decay had only wanted to be cradled in his knowledge; the unfixable hardly wants to be tinkered with, and scoffs at the presumption.

X remembered a joke about a Buddha on a mountaintop, dealing with a mountain-climbing seeker's questions as to the nature of the universe. The Buddha had put the seeker off with the explanation that the world rested on the back of an elephant, and then, a second time, with the explanation that the first elephant rested on the back of another. Finally, exasperated at the seeker's reappearance, the Buddha blurted, "It's elephants all the way down." An analogous cartoon depicted a group of scientists examining the cells of a dog on a laboratory slide, discovering that the dog's microscopic cellular structure consisted of innumerable tiny dogs. X bore through his mission here among the objects the knowledge that such allegories barely touched on the root truths. Everything was made of itself and everything else at once, dogs made of dogs but also cats, the Buddha himself the topmost elephant. X knew that if he'd tried to explain that he had x-ray vision, after being laughed out of the room he'd have been expected to give out the secret names, to expose the meat of everyday life. To explain what lay beneath the skin, as if life was an apple, with a bright thin beguiling peel overlaying something more actual.

Yet it was no apple. It was an onion. To discover that layers could be disassembled was to discover that nothing existed but layers, that the onion formed around itself only, all surface, all depth at once. And like an onion, what such encounters called forth were salty tears, which among other things served as the simplest reminder that your own surface was unsealed, penetrated from within by what you'd never be able to hide from another human gaze in a million years. X, Curator, had arrayed himself for the world and was naked nonetheless, like the artifacts he surveyed. What they told X anytime he stopped to listen was that he was merely a speck in a constellation, a lump formed in space and fallen to Earth for a brief, unspecified duration. To walk this planet and see what else was afoot was to be seen in turn, seen in time. Never mind the pretensions of archives, photography, curation. X's gaze froze nothing, because nothing was frozen. The heart was magma.

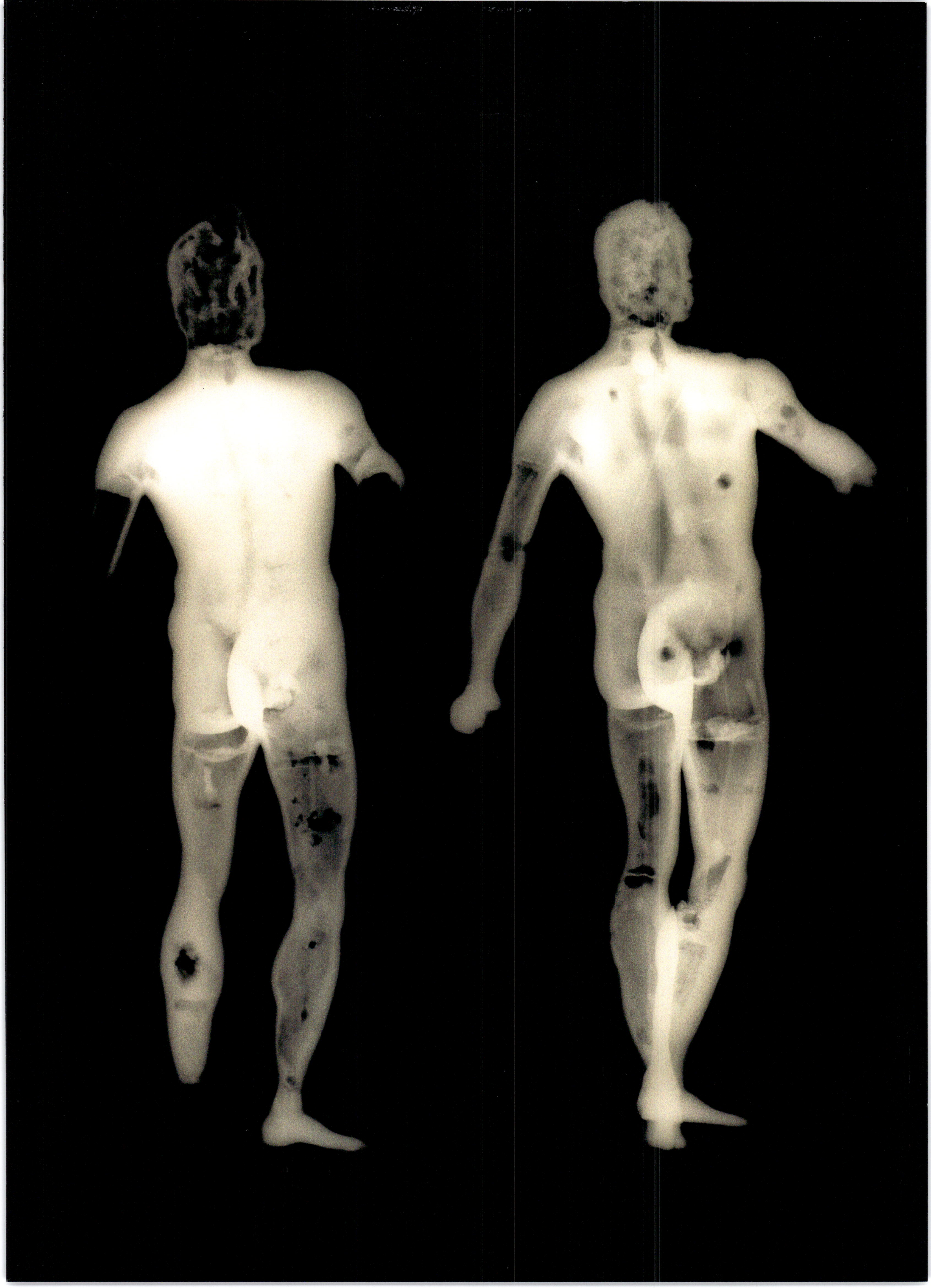

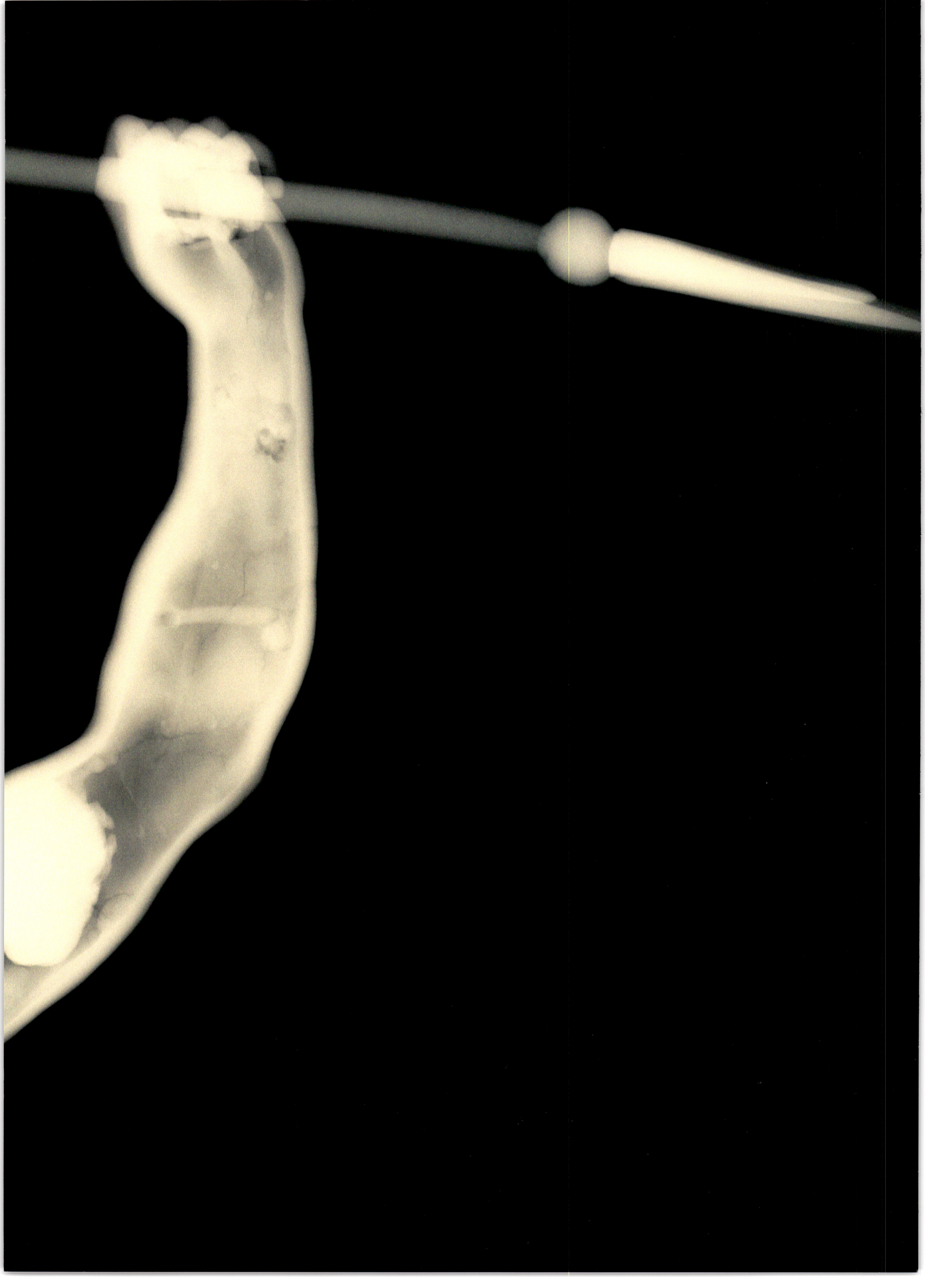

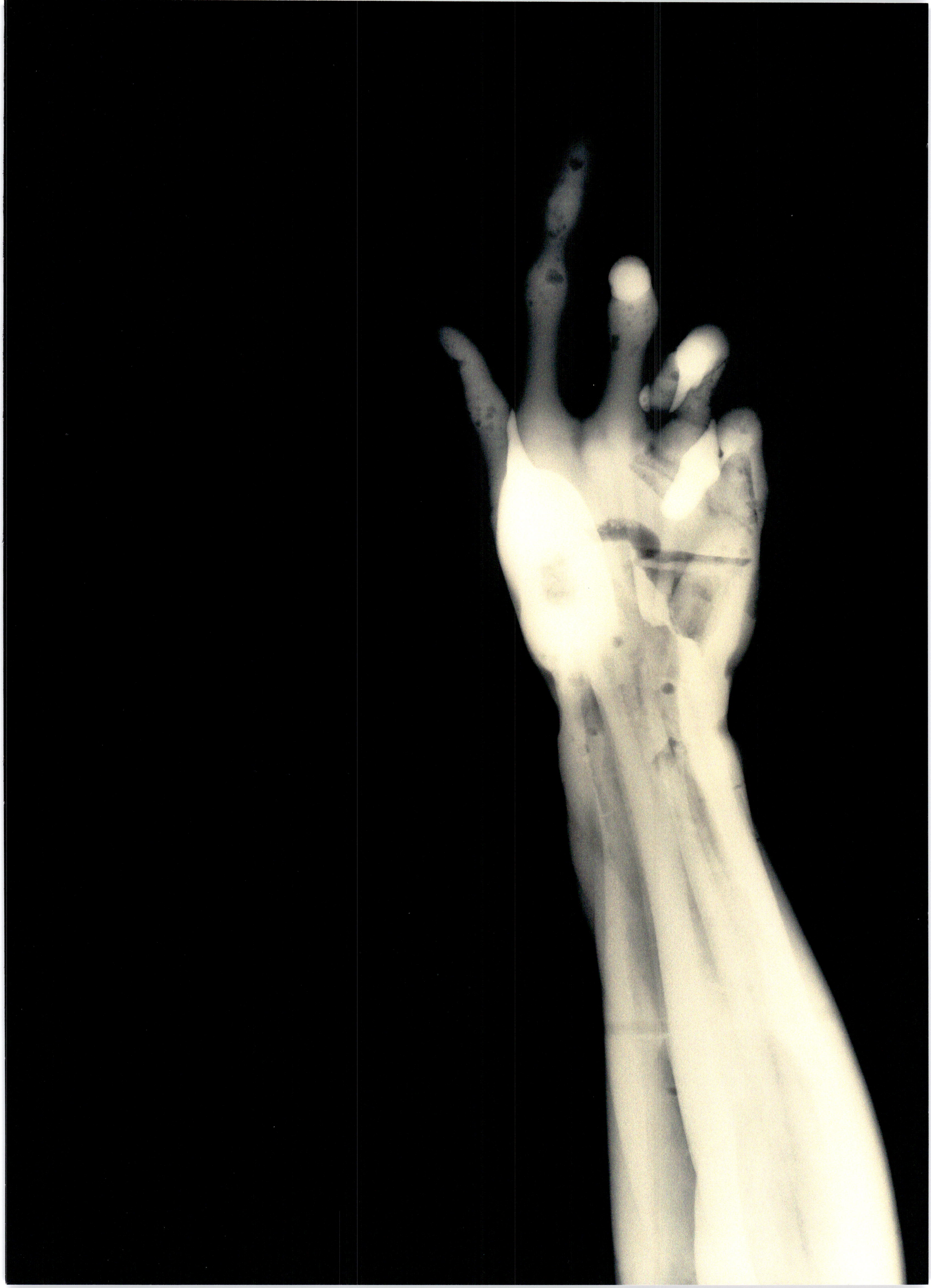

GM22 / GM21 / GM5 / GM20

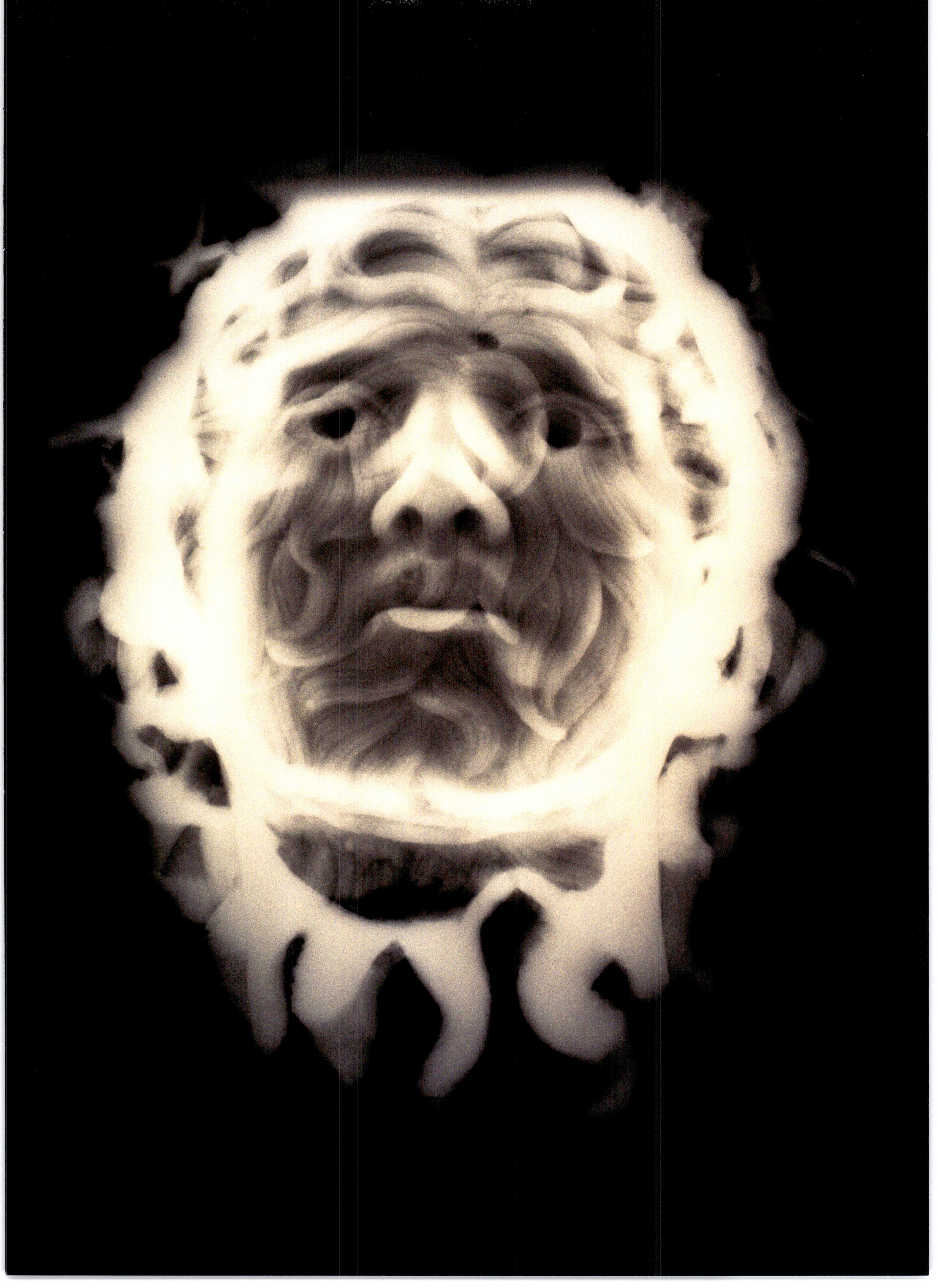

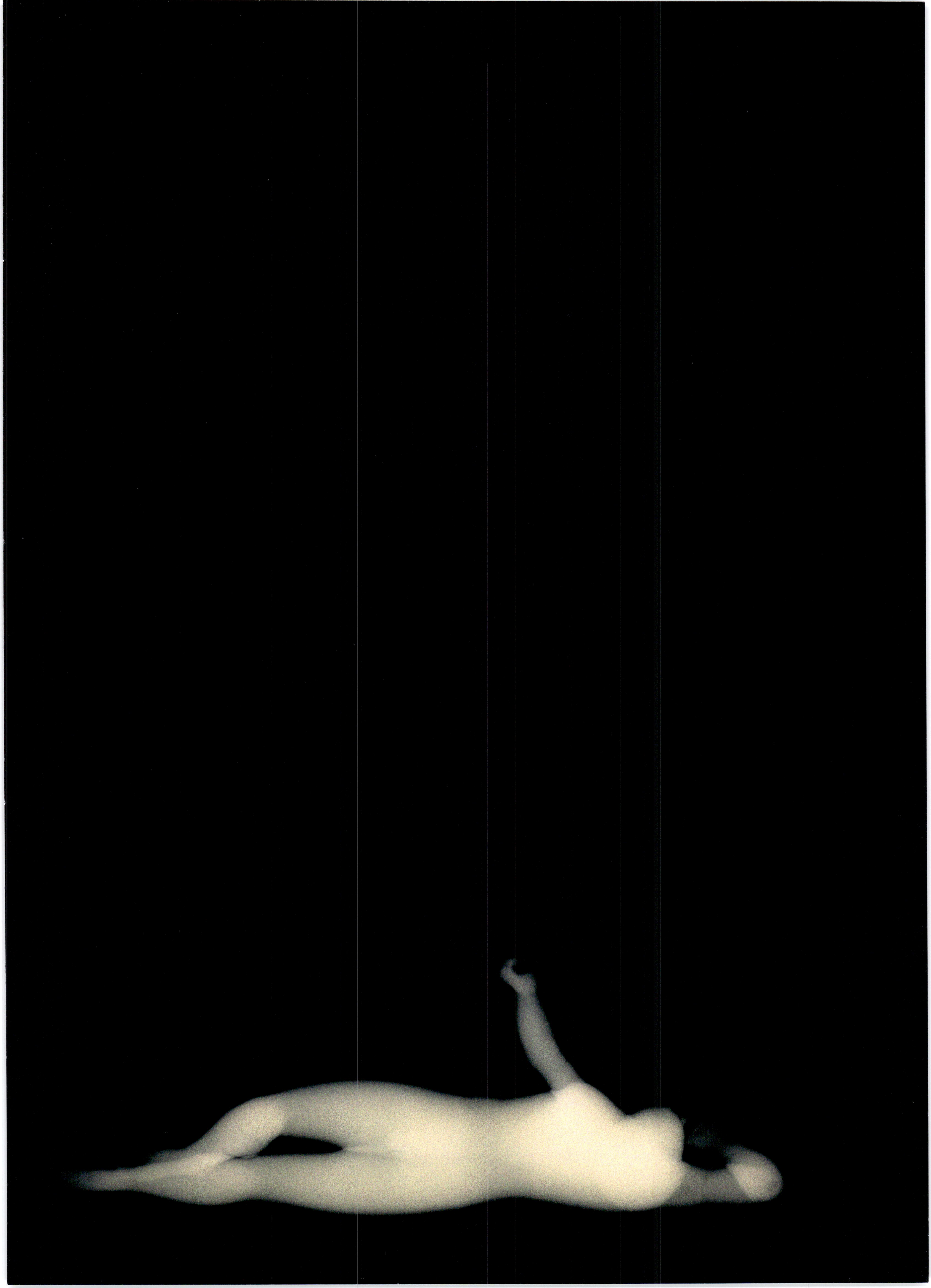

TRACE ELEMENTS AND CORE SAMPLES

My work in photography has been a long-term investigation into the aesthetics of entropy, and a parallel consideration of the dual processes of memory and excavation. The *History's Shadow* project provides for the continuation and expansion of these intertwined themes.

During a residency at the Getty Research Institute several years ago, I began to explore the idea of images that were created in the course of art preservation, where the realms of visual art and scientific research overlap each other. While photographing the Getty Museum's conservation departments, I became captivated by x-rays of art objects from the museum's permanent collections. The ghostly images of these x-rays seemed to surpass the potency of the original objects of art. These spectral renderings were like transmissions from the distant past, conveying messages across time, and connecting the contemporary viewer to the art impulse at the core of these ancient works.

Through the x-ray process, the artworks of origin become de-familiarized and de-contextualized, yet acutely alive and renewed, revivified. The shadow-worlds they occupy are informed by the black space surrounding the images, which in some instances becomes a vast nether world, and in others becomes the velvety ground of some kind of brain scan/portrait. The project's title of *History's Shadow* refers both to the literal images that the x-rays create as they are re-photographed, and to the metaphorical content informed by the past from which these objects derive.

Rendering three dimensions into two is at the heart of the photographic process. With the x-ray, this sense is compounded, since it maps both the inner and outer surfaces of its subject. The mysterious, archetypal images that result seem to encompass both an inner and an outer world, as the two-dimensional photograph brings us into a realm of indeterminate space, depth, and scale.

All of the x-rays I have photographed were elements of previously existing archives made for the distinct purpose of art conservation. It was the process of culling through many thousands of them, of excavating and bringing selected x-rays to light, which gave them their charge. To re-photograph these records, each was laid on a light box in a darkened room; the emanations of light were transmitted by long exposures onto color film, which was subsequently scanned and extensively manipulated.

The resulting images are both trace and index, as well as physical manifestations of chance, of subjectivity informed by the layering of different technologies and timeframes onto each other—the x-ray, the photograph, the scan; the filmic, the analogue, the digitized. At a critical juncture in the development of the photographic medium, when the digital seems about to subsume the analogue, these images extend the moment of the photographic instant, through the grafting and synthesis of multiple technologies and the depictions of the past filtered through them. In a sense, the project is an extended inquiry into some way of defining photography's essential qualities, its ways of constructing meaning.

The basic elements of *History's Shadow* are art, time, and technology. The x-ray-derived images serve as mediated expressions of the artists and artisans who created the original objects, however many centuries ago; as vestiges and indicators of the societies that produced these works; and as communications from the past, expressing immutable qualities that somehow remain constant over time. What do these works of art from past cultures have to teach us about our current point in human history, or about our relationship to the past, largely formed through archaeology and transmission of cultural objects across national borders? The x-ray provides a filter and a means (much as perception itself is both filter and means) to read the intrinsic properties of these works, the trace elements with which these objects are imbued. They encourage an understanding—made through feeling and art, as well as science and reason—that both spans and collapses time.

As with all photographs, these images are fragments; the x-ray, which in *History's Shadow* seems to slice through both material and time itself, furthers that connotation. The x-ray has historically been used for the structural examination of art and artifacts much as physicians examine bones and internal organs; it reveals losses, replacements, methods of construction, and internal trauma that may not be visible to the naked eye. These images, using x-rays as source material, are thus encrypted codes for the viewer to decipher. They make the invisible visible, and express through photographic means the shape-shifting nature of time itself, and the continuous presence of the past contained within us.

P L A T E S

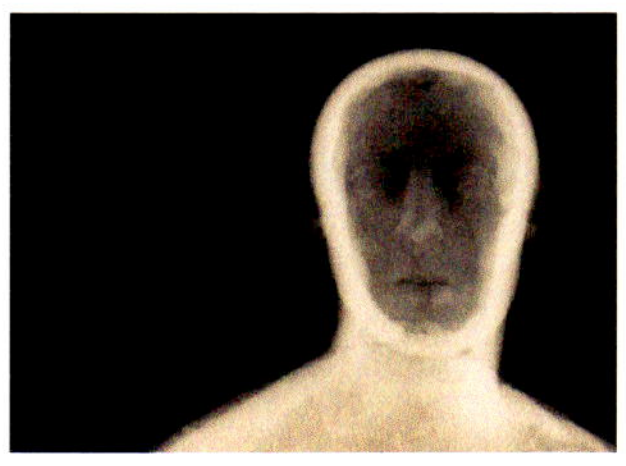
GM16

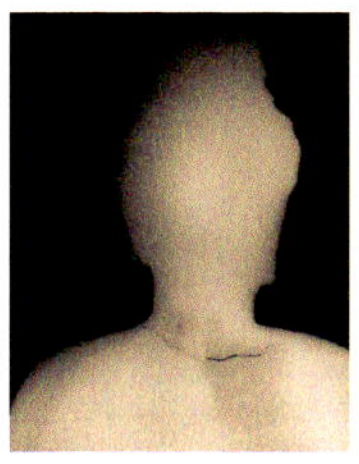
AB15

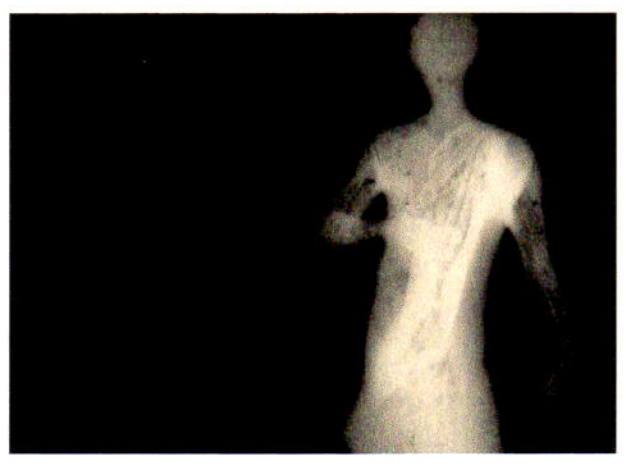
GM4

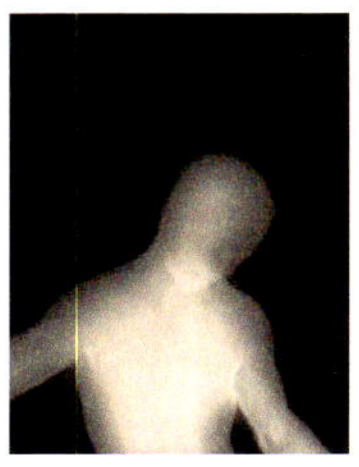
GM19

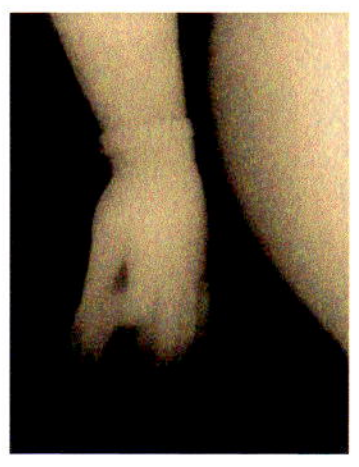
AB9

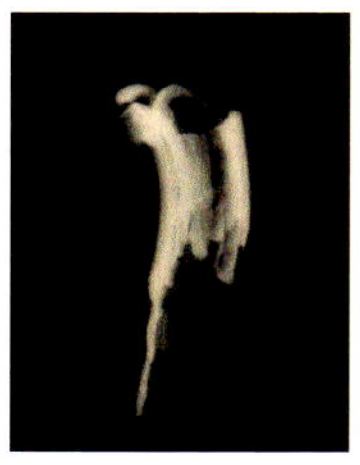
GM11

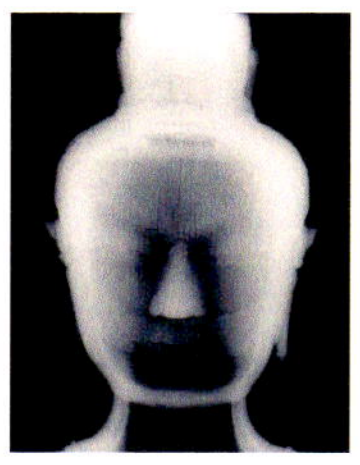
AB1

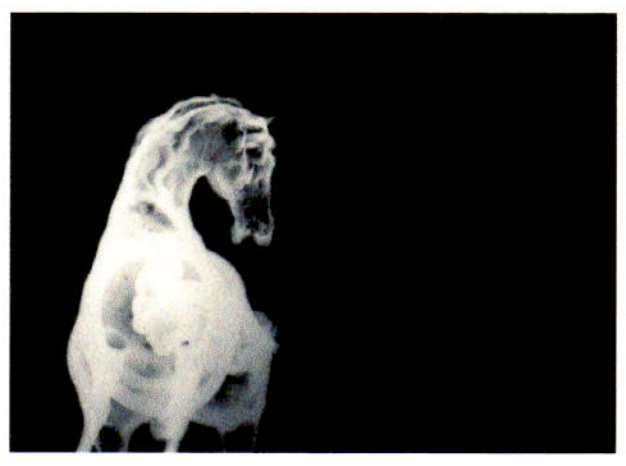
GM3

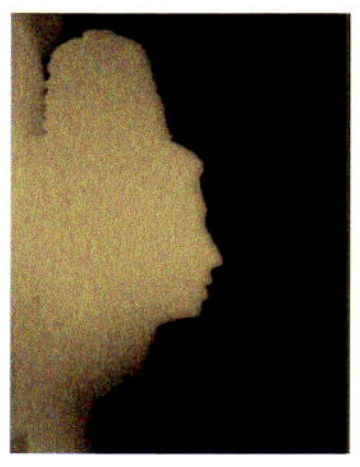
AB6

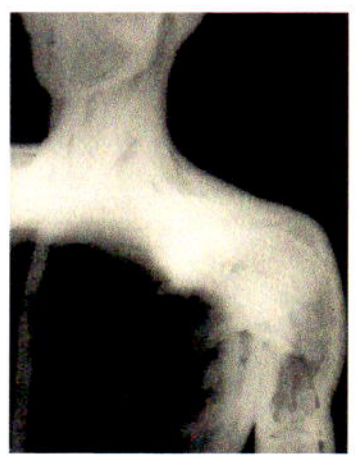
GM9

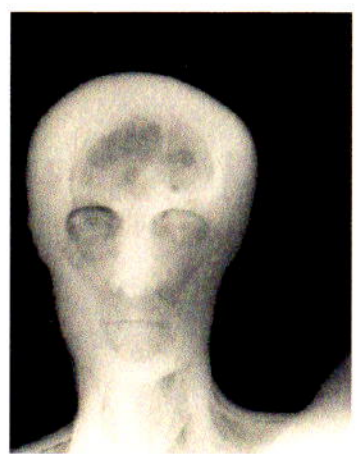
GM1

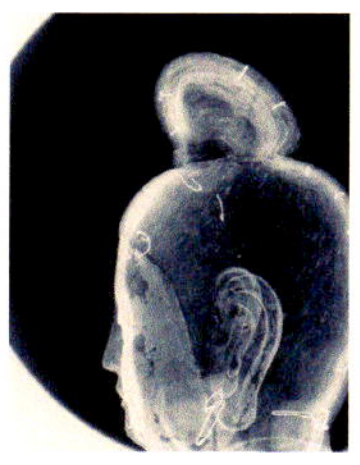
AB17

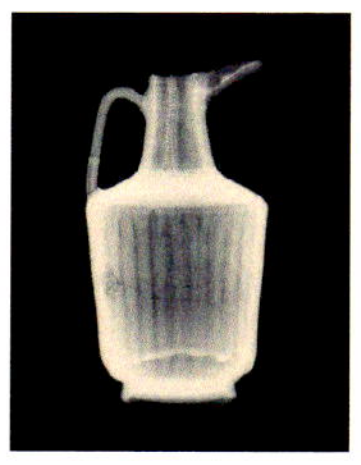
AV4

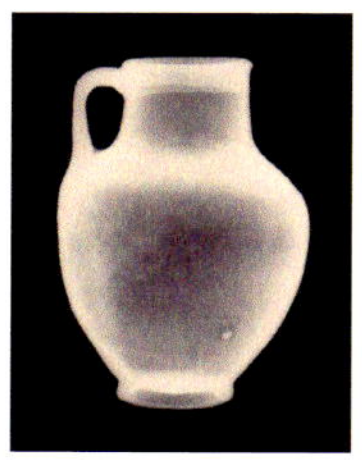
AV1

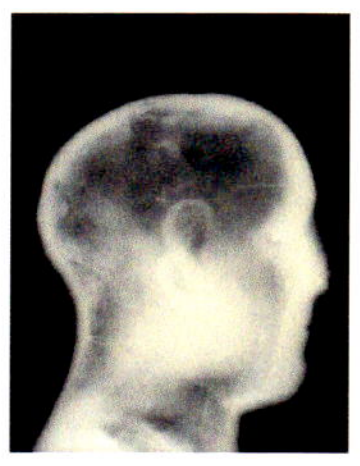
GM6

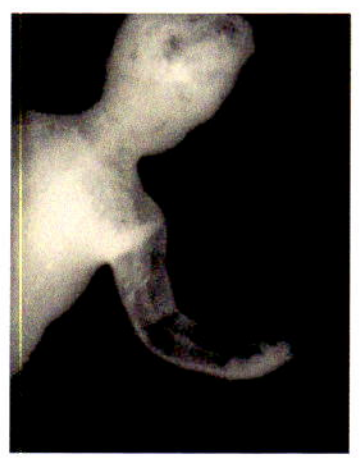
GM13

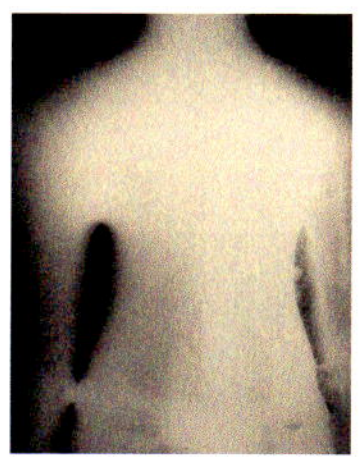
AB14

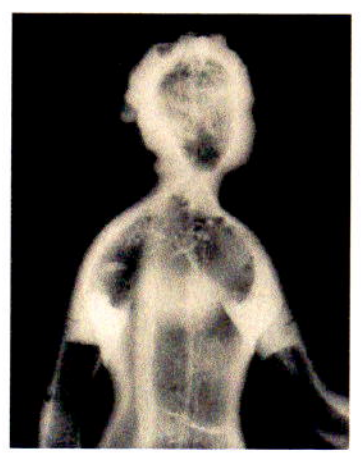
GM25

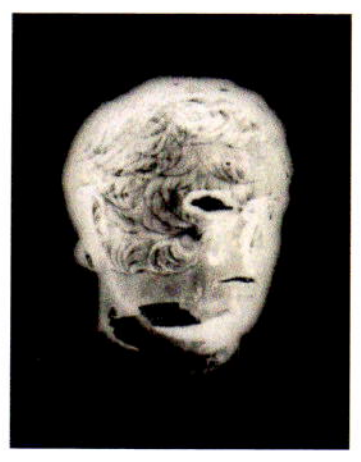
GM2

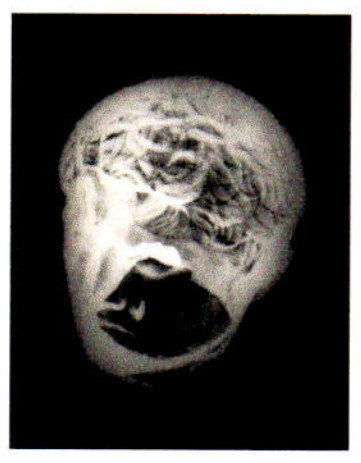
GM17

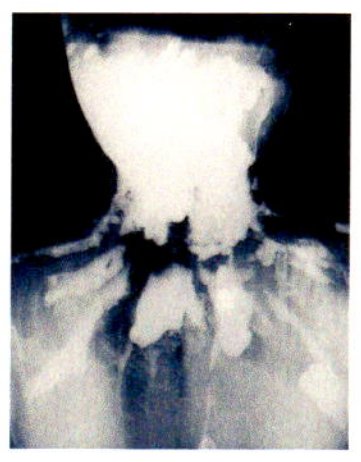
AB2

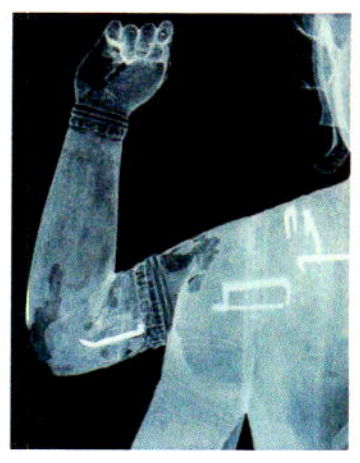
AB8a

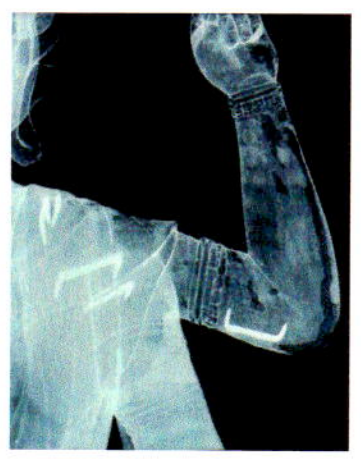
AB8b

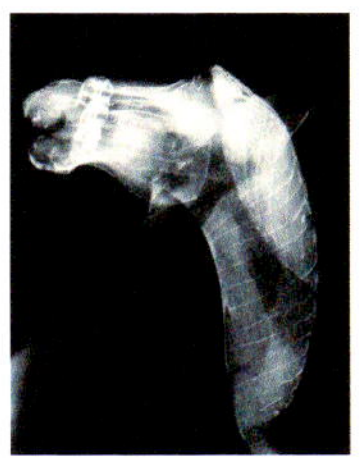
AB7

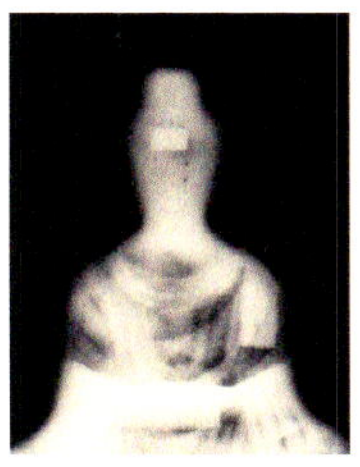
AB3

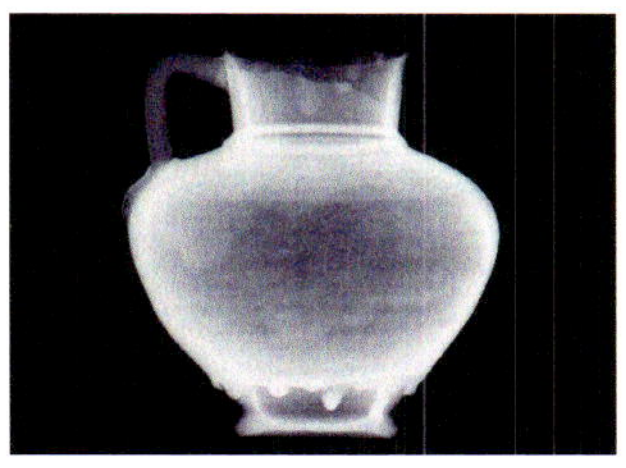
AV3

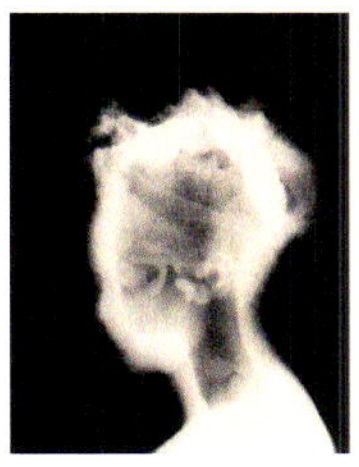
GM12

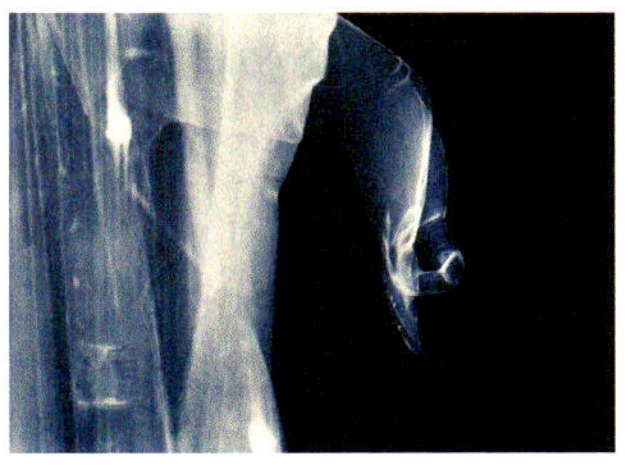
AB16

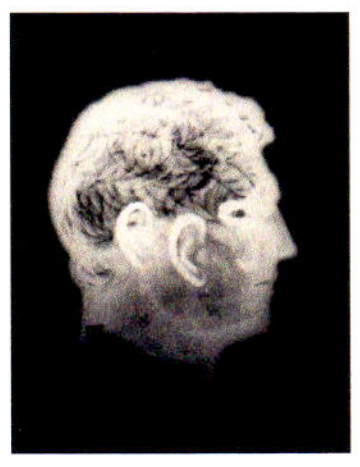
GM10

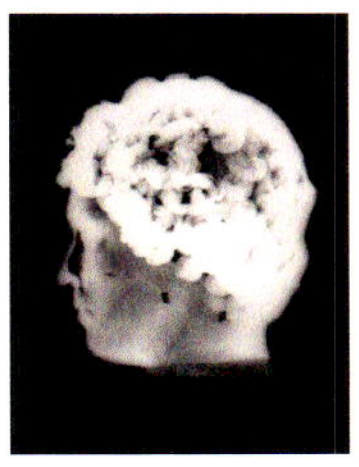
GM23

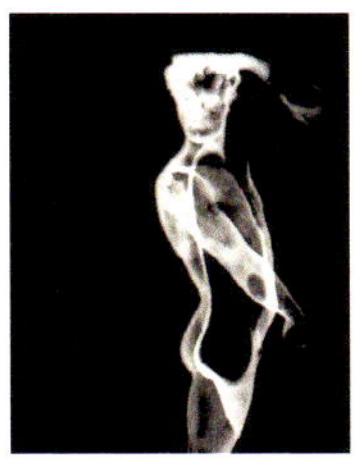
GM8

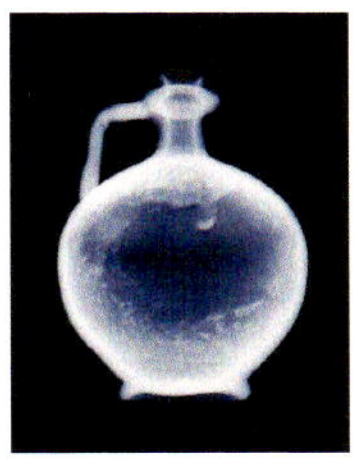
AV5

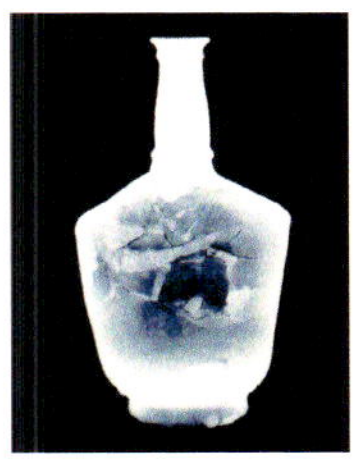
AV7

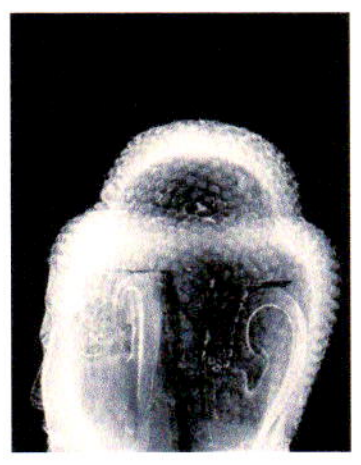
AB19

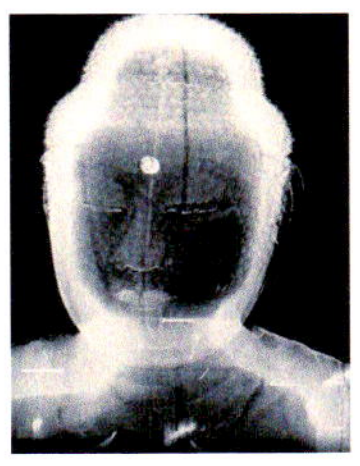
AB13

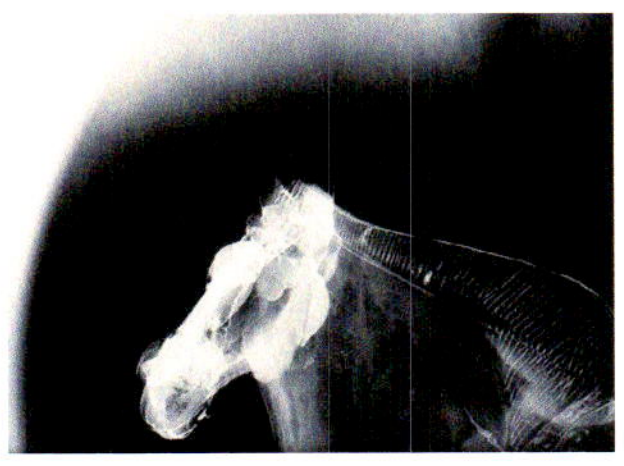
AB12

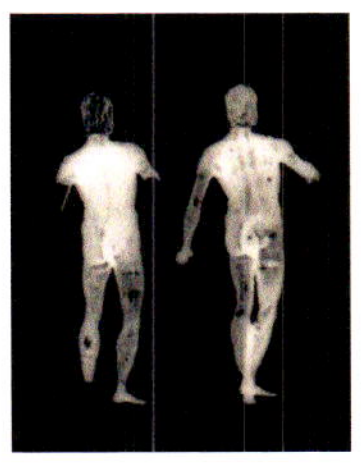
GM15

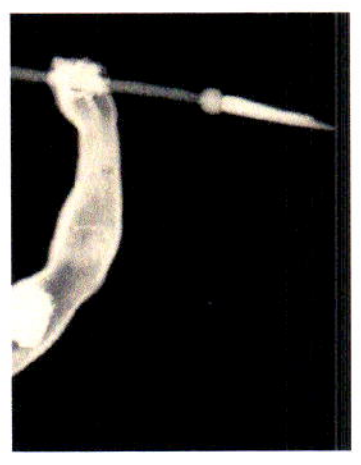
GM22

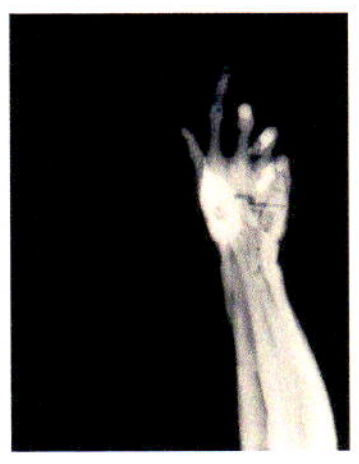
GM21

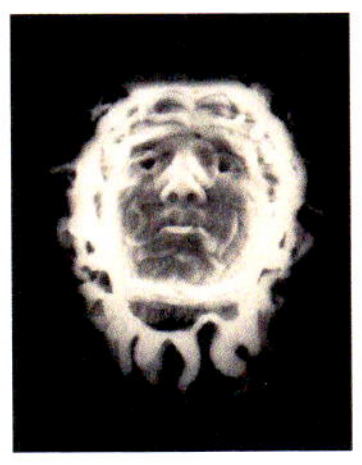
GM5

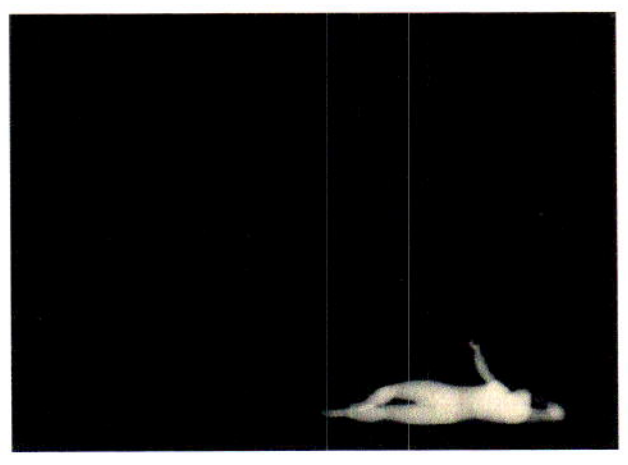
GM20

ACKNOWLEDGMENTS

The images in this volume of *History's Shadow* are derived from x-rays in the conservation departments of the Getty Museum and the Asian Art Museum of San Francisco. I am indebted to these institutions for granting me access and allowing me to engage with their radiography archives. This work was made possible due to their trust in my exploration and visual research.

I began *History's Shadow* while a Visiting Scholar at the Getty Research Institute. I am deeply appreciative of the support of the GRI, and the openness and assistance of all whom I encountered during my residency.

My thanks to those who have contributed to the development of this project:

Chris Pichler, for his friendship and his dedication to the art and craft of book publishing; Jonathan Lethem, for his brilliant and insightful story; Bob Aufuldish, for the remarkable design of this book; Lance Brewer, Tom Griscom, and Joseph Antonetti, for their skillful production assistance; Howard Stein and the Joy of Giving Something, for their ongoing support.

Colin Westerbeck, for organizing *Shadow and Dust*, the first solo exhibition to feature these images, which ran from August 31–December 31, 2010 at the California Museum of Photography, Riverside, CA.

From the Getty Center: David Bomford, Giacomo Chiari, Brian Considine, Thomas Gaehtgens, Tom Learner, Mark Leonard, Jeffrey Maisch, Jerry Podany, and Sabine Schlosser. From the Asian Art Museum of San Francisco: Jay Xu, Robin Groesbeck, and Aino Tolme.

The principals and staff of the galleries representing my work: Thomas Von Lintel, Dana Greenidge, and Nick Lamia, of the Von Lintel Gallery, New York; Cheryl Haines, Sasha Tierney, Monique Deschaines, and Sean Brimer, of the Haines Gallery, San Francisco; Ellen Miller, of the Ellen Miller Gallery, Boston; Holly Johnson, of the Holly Johnson Gallery, Dallas; Carrie Secrist and Natalie Popovic, of the Carrie Secrist Gallery, Chicago; and Elena Ochoa Foster, Antonio Sanz, and Dani Rivero Baughman, of the IvoryPress Gallery, Madrid.

From the Institute for Artist Management: Frank Evers, Matt Shonfeld, and Jasmin Chang.

My family, always in my heart: my wife Lynn and daughter Jessie; Howard, Eve, Miriam, Jennifer, Michael, Julia, and Joan.

. . . .

This book is dedicated to the memory of Larry Sultan.

David Maisel

Library of Congress Cataloging-in-Publication Data available.
ISBN: 978-1-59005-288-4

Manufactured in China.

Design by Bob Aufuldish, Aufuldish & Warinner

Nazraeli Press
2871 NE Alameda Street
Portland, OR 97212

www.nazraeli.com